THE **CHICKEN** ENCYCLOPEDIA

THE
CHICKEN
ENCYCLOPEDIA

AN ILLUSTRATED REFERENCE

GAIL DAMEROW
author of *Storey's Guide to Raising Chickens*

Storey Publishing

*The mission of Storey Publishing is to serve our customers by
publishing practical information that encourages
personal independence in harmony with the environment.*

Edited by Deborah Burns
Art direction and book design by Alethea Morrison
Text production by Sourena Parham

Cover and interior illustrations by © Bethany Caskey, except as noted on page 319
Interior photography credits appear on page 319

Indexed by Samantha Miller

Storey Publishing
210 MASS MoCA Way
North Adams, MA 01247
www.storey.com

Printed in China by R.R. Donnelley
10   9   8   7   6   5

Library of Congress Cataloging-in-Publication Data

Damerow, Gail.
   The chicken encyclopedia / by Gail Damerow.
      p. cm.
   Includes index.
   ISBN 978-1-60342-561-2 (pbk. : alk. paper)
   1. Chickens—Encyclopedias. I. Title.
SF481.3.D36 2012
636.503—dc23
                    2011026852

MY THANKS GO TO

Rebekah L. Boyd-Owens, editor of my earlier book *Storey's Guide to Raising Chickens,* third edition, for persistent and incisive questions that inspired this volume; Gene Morton, ornithologist and poultry enthusiast, for invaluable technical advice; Paul Kroll, APA and ABA judge, for help with fine-tuning some of the trickier definitions (although any misinterpretations remain entirely my own); Elaine Belanger, editor of *Backyard Poultry* magazine, for ongoing suggestions and encouragement; Allan Damerow, my husband and best pal, for research assistance and for indulging me in my passion for chickens.

COMMONLY USED ABBREVIATIONS

ABA	American Bantam Association
ACV	apple cider vinegar
AGB	American Game Bantam
AI	artificial insemination
ALBC	American Livestock Breeds Conservancy
AOC	All Other Colors; Any Other Color
AOCCL	All Other Comb Clean Leg
AOSB	All Other Standard Breeds
AOV	All Other Varieties; Any Other Variety
APA	American Poultry Association
BB	Best of Breed
BQ	breeder quality
BBR	black breasted red
BLR	blue laced red
CRD	chronic respiratory disease
DE	diatomaceous earth
EE	Easter Egger
FL	feather leg
GLW	golden laced Wyandotte

IB	infectious bronchitis
LS	light Sussex
MG	Modern Game
ND	Newcastle disease
NOP	National Organic Program
NSQ	not show quality
OEG	Old English Game
OEGB	Old English Game bantam
POL	point of lay
PQ	pet quality
RBC	Rare Breeds Canada
RFID	radio frequency identification
RIR	Rhode Island Red
RCCL	Rose Comb Clean Leg
SCCL	Single Comb Clean Leg
SLW	silver laced Wyandotte
SPPA	Society for the Preservation of Poultry Antiquities
SQ	show quality
VND	velogenic Newcastle disease

abdomen \ The belly, or the underside of a chicken's body from the end of the breastbone to the vent. Technically, because a chicken has no diaphragm separating its chest from its abdomen, its body cavity is more properly called the coelom. See page 19 for illustration.

abdominal capacity \ Total depth and width of the abdomen.

abdominal depth \ The distance between the pubic bones and the breastbone, indicating a hen's ability to hold a forming egg.

abdominal width \ The distance between the two pubic bones, indicating the amount of space available for an egg to pass through when being laid.
[Also called: width of body]

acariasis \ Infestation by mites.

addled \ Describes a rotten or otherwise inedible egg, typically one in which the embryo died during early incubation, such as when an egg was partially incubated, then abandoned.

aflatoxicosis \ A toxic reaction resulting from eating moldy grain containing aflatoxin, a compound produced by *Aspergillus flavus*, *A. parasiticus*, and *Penicillium puberulum* mold in grain. Aflatoxin is most likely to occur in grain that has been insect damaged, drought stressed, or cracked. Any grain that looks or smells moldy should not be fed to chickens.
[Also called: X disease]

age, determination of \ See box on page 8. *See also: longevity*

age of lay \ The age at which pullets begin laying eggs. Leghorns and similar lightweight breeds start laying at 18 to 22 weeks of age. Larger-bodied hens generally begin laying at 24 to 26 weeks of age.

aggressive chicken \ A chicken with an attitude. Mean individuals occasionally appear in nearly every breed and are more typically cocks than hens. Some breeds are characteristically aggressive; others commonly have aggressive individuals. Generally aggressive breeds include Cubalaya, Modern Game, and Old English Game. Breeds known for commonly having aggressive individuals include Aseel, Buckeye, Cornish, Faverolle, Rhode Island Red, Shamo, Sumatra, and Wyandotte. All sorts of methods have been put forth for taming an ornery rooster, but the safest course is to get rid of it before you, a family member, a neighbor, or a young child gets seriously injured.

abdominal depth

abdominal width

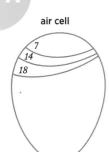

air cell

Relative size of air cell on the 7th, 14th, and 18th days of incubation.

aging (of meat) \ Allowing time for the muscles (meat) of a freshly killed chicken to relax and become tender. Aged chicken meat tastes better and is more tender than that of a chicken cooked or frozen a few hours after being killed. The muscle protein of a freshly killed chicken rapidly stiffens (as rigor mortis sets in), so unless the chicken is rushed to the cook pot, the meat will be tough if it is not aged before being cooked. The older the chicken, the longer its meat needs to age.

To age a freshly killed chicken, cool the meat and wrap it loosely. Set it in the refrigerator for at least one day, up to three days if it will be frozen or canned, or up to five days if it will be cooked fresh.

Agricultural Extension \ *See: Cooperative Extension System*

air cell \ The air space that develops at the large, round end of an egg between the inner and outer membranes just inside the shell. A freshly laid egg has no air cell. As the egg cools and its contents shrink, the inner shell membrane pulls away from the outer shell membrane, forming a pocket. The cell of a freshly laid cool egg is about ⅛ inch (0.3 cm) deep. As the egg ages, moisture evaporates from it, its contents continue to shrink, and the size of the air cell increases. Just how fast the cell grows depends on the porosity of the shell and

DETERMINING AGE

Exact age is not possible to determine, but a young bird may be distinguished easily from an old one based on the features listed below.

Feature	Young Chicken	Old Chicken
BODY SHAPE	Gangly	Round
LEGS	Smooth scales	Rough scales
SPURS	Small nubs	Long spurs*
BREASTBONE	Flexible	Rigid
MUSCLES	Soft	Firm
SKIN	Thin and translucent	Thick and tough

*All cocks and some hens have spurs; the longer the spur, the older the bird.

on the temperature and humidity under which the egg is stored.

During incubation the air cell increases rapidly, and its size at various stages of incubation may be used to determine whether or not the humidity within the incubator is optimal. If the air cell in an incubated egg is proportionately larger than is a typical cell after 7, 14, and 18 days of incubation, humidity is too low; if the cell is smaller, humidity is too high. *[Also called: air space]*

air cell gauge \ Gauge used for determining the depth of the air cell, one of the factors in grading an egg for quality (AA, A, B). The USDA official air cell gauge is designed to help the beginner learn to judge accurately the size of an air cell at a quick glance while candling. Experienced candlers occasionally use the gauge to verify the accuracy of their

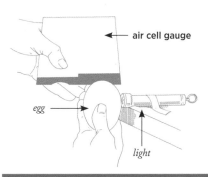

air cell gauge

egg

light

EGG QUALITY AND AIR CELL DEPTH

Quality	Depth
AA	⅛ inch (0.3 cm)
A	³⁄₁₆ inch (0.5 cm)
B	⅜ inch or larger (1 cm)

determinations. With the egg held large end upward, the gauge measures the depth of an air cell at the point of greatest distance between the top of the cell and an imaginary plane passing through the egg at the lower edge of the air cell where it touches the shell. *See also: candle*

air sacs \ Thin-walled pockets that, as part of the chicken's respiratory system, circulate air from the lungs throughout other parts of the chicken's body. The system of air sacs extends around the internal organs and into some bones, called pneumatic bones, that are hollow. Unlike the chicken's lungs, which are rigid, its air sacs are flexible. Of the nine air sacs, eight are paired, with one set attached to each lung. The single, largest air sac is shared by both lungs.

alarm call \ A sound a chicken uses to alert other chickens to potential danger. Different sounds are used to distinguish between a possible threat and immediate danger and between a predator in the air and one on the ground.

CAUTION CALL. A few quick notes briefly repeated, made by a chicken that sees (or thinks it sees) a predator in the distance. It is not particularly loud or insistent and doesn't last long unless the predator becomes a threat.

ALARM CACKLE. A more insistent caution call announcing the approach of an apparent predator on the ground or perhaps perched in a nearby tree or on a fence post. It consists of a brief series of short, sharp sounds followed by one loud, high-pitched sound. Other chickens may join the cackling while

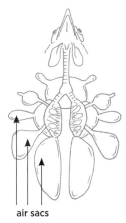

air sacs

stretching their necks to get a better look and moving around in an agitated way, as though not quite sure whether or where to run. These sounds increase in intensity the longer the assumed predator is in sight and may continue after the creature has gone.

AIR RAID. A loud warning cry made by a cock, or occasionally a hen, that spots the approach of a raptor. While making this sound the chicken looks up with one eye while flattening its head and tail in a crouch to make itself less conspicuous. Without looking up, the other chickens run for cover.   *See also: warning call*

albumen \ *See: egg white*

alektorophobia \ Fear of chickens (from the Greek words *alektor*, meaning cock, and *phobos*, meaning fear), usually triggered by some unpleasant incident involving a chicken. People with alektorophobia have an irrational fear of chickens and their parts (including a feather or an egg). They break out in a sweat or experience panic at the sight or nearness of a chicken, fearing it might peck or attack them. In an extreme case, just seeing a chicken in a photograph or movie can cause panic.

allele \ A variation of a gene (from the Greek word *allelos*, meaning "each other"); short for allelomorph (literally meaning "other form"). The combined effect of paired, unlike alleles can be either dominant or recessive.

all-in, all-out \ The management procedure of keeping laying hens or breeders for the most productive part of their lives, then disposing of the entire flock, thoroughly cleaning and sanitizing the facility, and bringing in a young replacement flock. The purpose of this method is to avoid introducing diseases that may occur when new chickens are brought into an existing flock or into unclean reused housing.

All Other Colors (AOC) \
A show designation for a catchall class that includes all colors of a breed or variety without a specific class of their own. [*Also called: Any Other Color*]

All Other Comb Clean Leg (AOCCL) \ One of the groupings into which the American Poultry Association organizes bantam breeds. It includes breeds that are not game breeds, have neither a single comb nor a rose comb, and are not feather-legged. This class includes Ameraucana, Araucana, Buckeye, Chantecler, Cornish, Crevecoeur, Cubalaya, Houdan, La Fleche, Malay, Polish, Shamo, Sicilian Buttercup, Sumatra, and Yokohama.

The American Bantam Association does not recognize Crevecoeur but additionally includes in this class Hamburg and Orloff.

All Other Standard Breeds (AOSB) \ One of six groupings into which the American Poultry Association organizes large chicken breeds. The breeds in this class are further subdivided into three groups: Game (Modern and Old English), Miscellaneous, and

Oriental. The Miscellaneous breeds are Ameraucana, Araucana, Frizzle, Naked Neck, and Sultan. The Oriental breeds are Aseel, Cubalaya, Malay, Phoenix, Shamo, Sumatra, and Yokohama.

All Other Varieties (AOV) \ A

show designation for a catchall class that includes all varieties of a breed without a specific class of their own.
[Also called: Any Other Variety]

alula (plural: alulae) \ A small,

sharp point in the middle of a chicken's wing, corresponding to the human thumb, covered with a few short, stiff feathers. Alula is the diminutive form of the Latin word *ala*, meaning "wing."
[Also called: bastard wing; false wing; spurious wing; thumb]

Ameraucana \ One of two breeds

that lay blue-shell eggs, developed from chickens imported from Chile. The name Ameraucana was derived by combining the words American and Araucana (the other breed that lays blue-shell eggs). Unlike the rumpless Araucana, the Ameraucana has a tail (although in Britain and Australia, rumpless and tailed are recognized as varieties of Araucana, not as separate breeds). It also sports a pea comb, as well as a beard and muffs. It comes in several color varieties, it may be either large or bantam, and the hens tend toward broodiness.

American Bantam Association (ABA) \ A national

organization established in 1914 to represent bantam breeders and promote all kinds of bantams. Its objectives are:
to encourage the breeding, exhibiting, and selling of purebred bantams

- to foster cooperation among breeders
- to sponsor national, semiannual, state, and special meets for the benefit of a greater bantam fancy
- to assemble and distribute information on the breeding, husbandry, and economic value of bantams

American class \ One of six

groupings into which the American Poultry Association organizes large chicken breeds. The 13 breeds in this class originated primarily in the United States and Canada and lay brown-shell eggs. They are Buckeye, Chantecler, Delaware, Dominique, Holland, Java, Jersey Giant, Lamona, New Hampshire, Plymouth Rock, Rhode Island Red, Rhode Island White, and Wyandotte.

alula

Ameraucana hen

American Game Bantam (AGB) \ A breed developed in the United States from the old-time pit game bantam. It is similar in appearance to the Old English Game bantam but slightly larger, and the cocks' tails feature a semi-heart-shaped curve that the Old English Game cock may or may not sport. Both Old English and American Games have a small single comb, but the American Game cockerel on exhibit must be dubbed to avoid disqualification, while the Old English cockerel may be shown undubbed.

The American Game comes in several color varieties, although not nearly as many as the Old English Game, and unlike the Old English is a true bantam, with no larger counterpart. American Game hens, like Old English hens, lay eggs with slightly tinted shells and brood easily. *See also: dub*

American Livestock Breeds Conservancy (ALBC) \

A national organization established in 1977 with the goal of ensuring the future of agriculture by conserving historic breeds and genetic diversity in poultry and other livestock. The Conservancy's programs include:

- research on breed population size, distribution, and genetic health
- research on breed characteristics
- gene banks to preserve genetic material from endangered breeds
- rescues of threatened populations
- education about genetic diversity and the role of livestock in sustainable agriculture
- technical support to a network of breeders, breed associations, and farmers

American Game Bantam hen

American Game Bantam rooster

American Poultry Association (APA) \ The oldest livestock organization in North America, established in 1873 with the primary objective of standardizing the many varieties of domestic fowl so their qualities might be fairly evaluated according to three guiding principles:

USEFUL TYPE. Within each breed the most useful type shall be the standard type

NEW BREEDS. No new breed may be recognized that cannot be readily distinguished from existing recognized breeds

COLOR VARIETIES. The recognition of color varieties in a breed is limited to a distinctive color or color pattern.

Other objectives include:
- to promote and protect the standard-bred poultry industry in all its phases
- to publish the *American Standard of Perfection* containing breed and variety descriptions for all recognized purebred fowl
- to encourage and protect poultry exhibitions as the show window of the industry, an education for both breeders and public, and a means of interesting future breeders in taking up poultry
- to assist, encourage, and help educate the junior poultry enthusiast to the sound and practical value of standard-bred poultry and pure breeding

American Standard of Perfection \ A periodically updated book, first published in 1874 as the *Standard of Excellence*, describing and depicting the 100-plus breeds currently recognized by the American Poultry Association. The organization's goal in establishing and illustrating the standard fowl was to stabilize the economic and commercial breeds to uniform size, shape, and color with good production and practicability, with the provision that ornamental breeds, including bantams, be attractive and productive and meet requirements of the standard breeder. Although this publication started out as the nation's premier sourcebook for the poultry industry, it has since narrowed its focus to exhibition. In addition to the breeds recognized in this volume, other breeds may be found in North America, and still more exist in the world.

amino acids \ Simple organic compounds occurring naturally in plant and animal tissues. Amino acids are the basic constituents of protein. Of the 22 amino acids that make up various kinds of protein, 14 are synthesized within a chicken's body. The remaining 8 (the essential amino acids) must be obtained through dietary protein.

ammonia \ A colorless gas that emits a characteristic pungent odor. The smell of ammonia coming from chicken housing means the bedding is too damp. When enough ammonia is emitted to make your eyes burn and your nose run, the ammonia level is high enough to increase your birds' susceptibility to respiratory disease. If the ammonia concentration gets so strong that the chickens' eyes become inflamed and watery and the birds develop jerky head movements, ammonia blindness may soon follow.

Where ammonia buildup becomes a problem, take measures to eliminate damp spots and condensation; frequently add fresh litter or decrease the number of birds housed; aerate the litter often by loosening and turning it; and improve ventilation to remove excess moisture from the air. Periodically applying ground rock phosphate or ground dolomitic limestone also helps keep litter dry and improves the fertilizer value of droppings-laden litter. For an average-size backyard coop, apply 1 pound (0.45 kg) per week stirred into litter or 2 pounds (0.9 kg) per week scattered over the droppings pit.

amprolium \ A commonly used coccidiostat available under several different brand names.

anatomy \
See box on pages 18 and 19. *For egg anatomy, see page* 54.

Ancona single comb hen, bantam size

Ancona \ An Italian breed named after its town of origin. The Ancona is so similar in type to the Leghorn that it is sometimes called a mottled Leghorn, from the white speckles in its black feathers. The Ancona may be either large or bantam and comes in two comb varieties — rose comb and single comb. The hen's comb typically flops to one side. The hens lay white-shell eggs and seldom brood.

Andalusian \ A breed originating in Spain and named after the region where it originated. It comes in only one color, blue, and is sometimes called the blue Andalusian, although only 50 percent of the offspring of a blue cock mated to a blue hen will be blue; half the remainder will be black and the other half splashed. Andalusians may be either large or bantam. They sport a single comb, and the hen's comb typically flops to one side. The hens lay chalky white-shell eggs and seldom brood.

Andalusian blue \
See: Andalusian; blue

anemia \ A condition in which the blood is deficient in quantity (blood loss) or quality (low hemoglobin, red blood cell count, or both), caused by dietary iron deficiency, internal parasites (worms), external parasites (mites or lice), coccidiosis, or some infectious disease — notably infectious anemia. Signs include pale skin and mucous membranes, loss of energy, and loss of weight despite ravenous appetite. *[Also called: going light]* See also: *infectious anemia*

angel wing \ *See: slipped wing*

animal protein \ Protein derived from animal sources for the purpose of furnishing the eight essential amino acids that are not synthesized within a chicken's body. Animal protein in commercially prepared feeds can be problematic if it is derived from diseased animals or livestock that has been fed antibiotics and other undesirable drugs. Chickens that forage freely obtain animal protein from bugs, worms, and other small critters.

ankle \ The hock joint, often mistakenly called the knee.

anthelmintic \ Any preparation used for deworming, from the Greek words *anti*, meaning "against," and *helmins*, meaning "worms." *See also: dewormer*

antibiotic \ A medicine used to destroy or inhibit the growth of bacteria and other microorganisms. Although antibiotics have been fed (controversially) to commercially grown broilers to stimulate growth and feed-conversion efficiency, they are not used in feeds sold for backyard flocks.

antibody \ A natural substance in the blood that recognizes and destroys foreign invaders and causes an immune response to vaccination or infection. A chick acquires disease protection through antibodies obtained from the mother hen via substances in the egg, primarily in the yolk. For about four weeks after hatching, therefore, a chick is immune to any disease for which the mother hen carries a high level of antibodies. Since antibodies attach to specific antigens, the antibody against one disease offers no protection against any other disease.

anticoccidial \ One of many drugs used to prevent or treat coccidiosis. A drug used for prevention is a coccidiostat; a drug used for treatment is a coccidiocide.

anticoccidial vaccine \ A commercial vaccine administered to chicks to stimulate immunity by introducing a low-level infection from the species of coccidia most likely to be encountered in their environment. Immunity is subsequently enhanced by exposure to the developing life cycles of both the coccidial strains present in the vaccine and those naturally occurring in the environment. Many hatcheries offer the option

Andalusian hen

of having chicks vaccinated, which produces lifetime protection against coccidiosis, provided the chicks are never fed medicated rations, which would neutralize the vaccine.

antigen \ A toxic or other foreign substance that triggers an immune response, especially by stimulating the production of antibodies.

antitoxin \ An antibody that is artificially introduced through injection and that neutralizes toxins produced by bacteria. Botulism, for example, is caused by bacteria that produce toxins. An injection of botulism antitoxin may be used to treat the disease by inducing immediate, though temporary, immunity.

Antwerp Belgian \ An alternative name for the Belgian Bearded d'Anvers bantam, named after the Belgium municipality of Antwerp (which the French call Anvers). *See also: Belgian Bearded d'Anvers*

Any Other Color (AOC)\
A show designation for a catchall class that includes all colors of a breed or variety without a specific class of their own. *[Also called: All Other Colors]*

Any Other Variety (AOV) \
A show designation for a catchall class that includes all varieties of a breed without a specific class of their own. *[Also called: All Other Varieties]*

APA–ABA Youth Program \
A national organization sponsored jointly by the American Poultry Association and the American Bantam Association to educate young folks five years or older about poultry by promoting opportunities to practice showmanship and by encouraging members to become involved with the poultry fancy in general.

Appenzeller \ A chicken developed in the Appenzell canton of Switzerland. Appenzellers are of two distinct breeds: Spitzhauben (which has found its way to North America) and Barthühner (which has not). *See also: Barthühner; Spitzhauben*

apple cider vinegar (ACV) \
See: vinegar

approved for poultry \
Ascertained by the Food and Drug Administration to be generally safe for use around chickens and other poultry and for which a withdrawal period (if applicable) has been established. Medications, dewormers, paraciticides, pesticides, and similar products that have not been approved for poultry must be used with caution, preferably under veterinary supervision.

apron \ A fencing technique to keep digging predators such as raccoons and foxes from burrowing into the poultry yard. It is created by attaching a 12-inch (30 cm)-wide length of wire mesh to the bottom of the fence, perpendicular to the fence and extending away from the yard. On a portable run, the apron sits on top of the ground; on a fixed run, it is usually shallowly buried. To prevent soil moisture from rapidly rusting the apron, use vinyl-coated wire or brush the mesh with roofing tar. Cut and lift the sod along the outside of the fence line, and clip or lash the apron to the bottom of the fence. Spread the apron horizontally along the ground and replace the sod on top. The apron will get matted into the grass roots to create a barrier that discourages digging. *[Also called: fence skirt]* \ A saddle.

aragonite \ Calcium carbonate derived from seashells and used as a supplement for laying hens to ensure strong eggshells.

Araucana \ One of two breeds that lay blue-shell eggs (the other being the Ameraucana), developed from chickens imported from Chile and named after Indians living in Chile's Gulf of Arauco. Araucanas have a pea comb, come in several color varieties, and may be large or bantam. Their most distinguishing physical features are their lack of a tail, reducing fertility (because a tail pulls feathers away from the vent during mating), and their spectacular ear tufts, a trait associated with a lethal gene that can cause the early death of chicks. Araucana hens tend toward broodiness. *See also: ear tuft; rumplessness*

Araucana hen and chicks

ANATOMY

Capably discussing various aspects of a chicken requires being able to identify its many parts.

See pages 115 and 293 for details of wing anatomy. See page 44 for butchering anatomy.

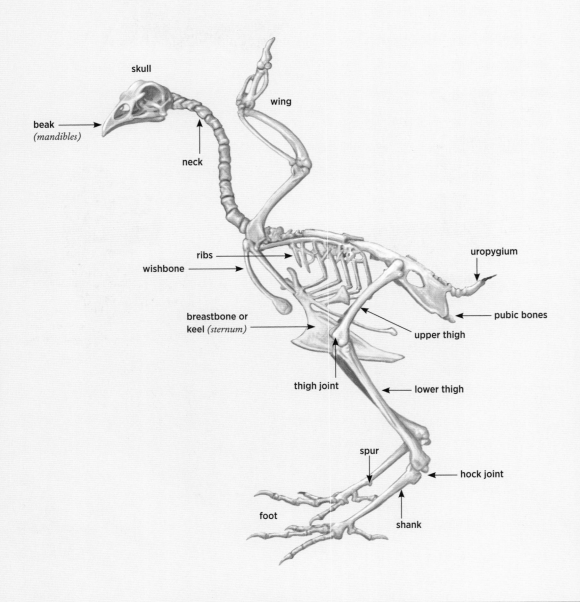

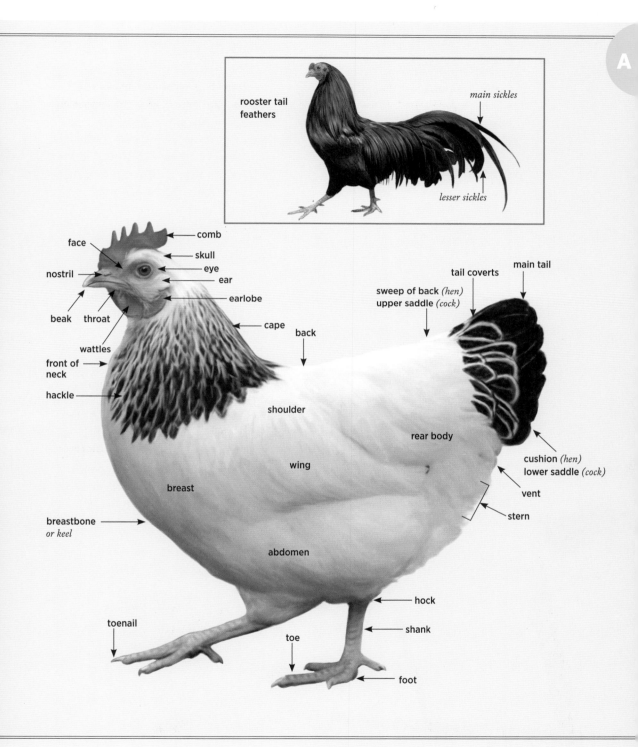

rooster tail
feathers

main sickles

lesser sickles

comb

face

skull

eye

nostril

ear

earlobe

beak   throat

cape   back

wattles

front of
neck

hackle

shoulder

tail coverts   main tail

sweep of back *(hen)*
upper saddle *(cock)*

rear body

wing

cushion *(hen)*
lower saddle *(cock)*

breast

vent

breastbone
or keel

stern

abdomen

hock

shank

toenail

toe

foot

ark \ A small, floorless, portable chicken shelter, usually A-frame in shape, periodically moved around the lawn or garden to provide fresh forage and a clean environment for the inhabitants.
[Also called: chicken tractor]

arthritis \ Inflammation of the joint and surrounding tissue. The most common type of avian arthritis is staphylococcic arthritis, caused by the *Staphylococcus aureus* bacteria, common in poultry environments. The bacteria enter a chicken through an open wound and may cause food poisoning in humans if an infected chicken is handled in an unsanitary manner during or following butchering. Prevention involves avoiding injuries by providing safe, uncrowded housing.
[Also called: joint ill]

artificial incubation \ Hatching eggs in an incubator, rather than under a hen.   *See also: incubator*

artificial insemination (AI) \ The injection of semen into a hen other than by natural mating with a cock. Artificial insemination is used by breeders of Cornish chickens (which have trouble mating naturally because of their heavy muscling and wide-apart legs), by exhibitors who wish to keep their hens in show condition or have a top cock with a leg injury that precludes mating, and by owners of valuable cocks that tend to be shy or otherwise low in sex drive.

ascaridiasis \ Infection by ascarids.

ascarids \ Large, long, yellowish-white parasitic nematode worms of the family Ascaridae. These worms invade the chicken's intestine, causing a pale head, droopiness, weight loss, diarrhea, and death. Most chickens become resistant to ascarids by three months of age.
[Also called: large roundworms]

ascites \ (pronounced a-site-eez) An accumulation of yellowish or bloody fluid in the abdominal cavity, generally occurring in fast-growing broilers as a result of heart failure. Prevention includes avoiding stress caused by crowding, insufficient numbers of feeders and

ark
or chicken tractor

drinkers, excessive temperatures (hot or cold), ammonia fumes arising from damp bedding, and inadequate ventilation. *[Also called: broiler ascites; dropsy; waterbelly]*

Aseel \ A muscular game breed developed more than three thousand years ago on the Indian subcontinent. The name Aseel — an Arabic word meaning "trueborn" or "purebred" — was given to this indigenous breed while northern India was under Muslim rule. Aseels are known for their short, tight plumage; upright stance; and square shanks, as well as the cocks' inherent disposition to engage in vicious fights if not housed separately. The Aseel has a pea comb, lacks wattles, and comes in a few color varieties. A bantam version is known in Europe but not in North America. The eggs have a tinted shell color, and hens quite typically go broody.

as hatched \ *See: straight run*

Asiatic class \ One of six groupings into which the American Poultry Association organizes large chicken breeds. The breeds in this class, originating primarily in China, are Brahma, Cochin, and Langshan. All three breeds are feather-legged and tend to be quite large.

Asil \ Aseel.

aspergillosis \ Brooder pneumonia.

Aseel hen

Aseel rooster

Australorp
hen

Australorp \ A breed developed in Australia, primarily from the black Orpington, as a dual-purpose farmstead chicken. The breed name was derived by combining the words Australia and Orpington. The Australorp has a single comb, comes only in black, and may be either large or bantam. The hens are good layers of brown-shell eggs and have a quiet disposition that makes them first-rate brooding hens.

automatic door closer \ A mechanical device that is regulated either by a timer or by sunlight to close a pop-hole door in the evening and open it in the morning. Since chickens actively forage at dusk, the door must be set so it won't close until they are all safely inside. In the morning the door might be set to open after the hens have finished laying, to prevent the ritual hunt for eggs laid outside the shelter.
[Also called: door keeper]

automatic turner \ A mechanical device that periodically turns eggs in an incubator so it doesn't have to be done manually.

autosexing\ Revealing clearly distinct characteristics at hatch, by which cockerels may easily be distinguished from pullets within a single breed. The term autosexing was developed to differentiate between sex-link purebreds and sex-link crossbreds. Autosexing breeds include the barred Plymouth Rocks and other barred varieties, light-brown and silver varieties of Leghorn, Norwegian Jaerhon, and the Crele variety of Penedesenca.

Many autosexing breeds were developed through crosses with a barred Plymouth Rock, combining the two names in the resulting breed: Ancobar (Ancona crossed with Barred Rock), Cambar (Campine), Cobar (Cochin), Legbar (Leghorn), Rhodebar (Rhode Island Red), Welbar (Welsumer), and so forth. At hatch the cockerels are lighter in color than the pullets.

Autosex breeds were developed primarily in Cambridge, England, during the 1930s, and most are now either rare or extinct. In the United States the autosexing California Gray was developed in the 1930s by crossing white Leghorns with Barred Plymouth Rocks, but it, too, is now quite rare.   *See also: sex-link*

autosomal chromosome \

Any chromosome that is not a sex chromosome. Of a chicken's 39 pairs, 38 are autosomal.

avian \ Pertaining to birds, from the Latin word *avis*, meaning "bird."

avian influenza \ *See: bird flu*

avidin \ A protein in raw eggs. Among the various proteins, avidin is significant because it ties up the vitamin biotin as part of an egg's defenses against bacteria, since most bacteria can't grow without biotin. Dogs and cats are sensitive to the effects of avidin and therefore should not be fed raw eggs routinely. A human, however, would have to eat two dozen raw eggs a day to be affected. Cooking eggs inactivates avidin.

axial feather \ The single short feather growing between and separating the wing's primary and secondary feathers. When you spread out a chicken's wing, the axial feather is roughly in the middle and indicates where the primaries end and the secondaries begin.
[Also called: key feather]

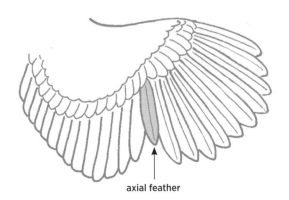

axial feather

bandette

back \ The top section of a chicken's body from the base of the neck to the base of the tail, including the cape and saddle. The length, shape, and slope of the back vary with the breed and are important distinguishing features. *See also: cape; saddle*

balanced ration \ A ration that provides all the nutrition a chicken needs, which varies with season, stage of growth, and production. Improper nutrition — resulting, for example, from feeding a ration that contains insufficient calcium, carbohydrate, or protein or perhaps from feeding too little of a balanced ration — can cause a reduction in laying. Cold weather increases a chicken's need for carbohydrates; failing to adjust the ration accordingly may result in low production. Common causes of imbalanced ration are neglecting to furnish layers with a free-choice calcium supplement and feeding hens too much scratch.

band \ A wing band, leg band, or bandette. \ To apply a wing band, leg band, or bandette.

bandette \ A numbered plastic spiral applied to a chicken's leg for identification purposes. Bandettes generally come in lots of 25, all of one size and color, and are sequentially numbered. Different colors might be used to denote different years or family lines. Breeders who double-mate typically use blue bandettes for the male line and red (or pink) ones for the female line. Since bandettes eventually get brittle and break, a good plan is to band both legs with bandettes of the same color and number.

bantam \ A small chicken, about one-fourth to one-fifth as heavy as a large-size chicken and generally weighing 2 pounds (0.9 kg) or less. Not all bantams have the same ancestry as the large version of the same breed; some were developed from entirely different bloodlines to look similar, only smaller.

Although nearly every breed and variety of large chicken has a bantam version, some bantams have no large counterpart. Those that do not are called true bantams, while those that do are called miniatures. But they are not exact miniatures — the size of their head, tail, wings, feathers, and eggs is larger than if they were perfect miniatures.

Bantam Standard \ A book published by the American Bantam Association describing each of the bantam breeds recognized by that organization. Its information doesn't always coincide with the American Poultry Association's *Standard of Perfection* as to breeds listed, exact breed names, and descriptions.

banty \ Short for bantam.

barbs \ The slender filaments of a feather extending from the shaft and bearing barbicels with interlocking hooks. The hooks fasten together to form the smooth web characteristic of the feathers of most breeds, except the Silkie and the silkie feathered Hedamora.

Barnevelder \ A dual-purpose breed developed in the municipality of Barneveld in the middle of the Netherlands, an area with a strong poultry industry. The Barnevelder has a single comb, comes in a few color varieties (although only one, double laced partridge, is recognized in North America), and has a bantam counterpart in Europe but not in North America. The hens lay well, producing large eggs with dark brown shells, but are not particularly inclined to brood.

barny \ Short for a barnyard chicken. \ When capitalized, short for Barnevelder.

barnyard chicken \ A chicken of mixed heritage.

barred, barring \ A color pattern in feathers consisting of alternate crosswise stripes of two distinct colors. The stripes are of equal width in Barred Plymouth Rocks, unequal in Campines, Dominiques, and Hollands. The irregular stripes of gold and silver penciled Hamburg females may also properly be called barring. \ A color defect in a black feather created by the appearance of crosswise purple stripes.

barred feather

Barnevelder hen

Barthühner
hen

beak \ The hard, protruding portion of a bird's mouth, consisting of an upper and a lower mandible that serve as the bird's jaws.

beak beating \ A feeding behavior in which the chicken picks up a bit of food in its beak and rubs it on the ground to break off pieces small enough to swallow.

beak conditioning \ An industry euphemism for debeaking. *See also: debeak*

beaking out \ The practice of using the beak to scoop feed from a trough onto the ground. Feed on the ground is usually wasted because it gets trampled into the dirt or bedding and covered with droppings. To discourage beaking out, use a feeder or trough with a rolled or bent-in edge and adjust the hopper to the height of the chickens' backs. When raising chicks, fill chick-size trough feeders only two-thirds full and continue adjusting feeder height as the birds grow to maturity.
[Also called: billing out]

beak wiping \ A grooming behavior in which the chicken wipes its beak along the ground. This activity serves three purposes: it cleans the beak, sharpens the beak, and keeps the ends from growing too long or out of balance.

Barthühner \ One of Switzerland's two Appenzeller breeds, little known outside its native country. Barthühner means "bearded chicken" in German, the primary language of the region. This breed is quite chunky in appearance, sports a rose comb in addition to a beard and muffs, comes in a few color varieties, and may be large or bantam. The hens lay white-shell eggs and may brood.

bastard wing \ Alula. The term is used in the old sense of the word, meaning "false," because it looks like a tiny false wing growing out of the regular wing. *See also: alula*

'bator \ Short for incubator.

beard \ A dense clump of small feathers attached to the upper throat, just under the beak, always in association

B

with a muff. Some breeds, such as Belgian Bearded d'Anvers and Belgian Bearded d'Uccle, are clearly identified as bearded. Other bearded breeds include Ameraucana, Crevecoeur, Faverolle, and Houdan. Some breeds, notably Polish and Silkie, come in bearded and non-bearded varieties.

Bearded d'Anvers \ *See: Belgian Bearded d'Anvers*

Bearded d'Uccle \ *See: Belgian Bearded d'Uccle*

bedding \ Any material — such as chopped straw, wood shavings (except cedar, which is toxic to chickens),

shredded paper, or dried grass clippings — scattered on the floor of a chicken coop to absorb moisture and droppings, cushion the birds' feet, help minimize breast blisters and other injuries to heavy breeds, and control temperature by insulating the ground.
[Also called: litter]

beetle brow \ A prominent forehead that projects out over the eyes — typical of Brahmas, Malays, and Shamos — giving the chicken the often undeserved appearance of being mean or sinister.

beard

B

beetle brow

BEAK TRIMMING

Sometimes the length of the beak's two mandibles must be adjusted so the chicken can properly peck and preen. Like fingernails a beak grows continuously, but unlike fingernails it normally wears down as fast as it grows, aided by the activity of beak wiping. When a chicken lacks opportunities to keep its beak worn down (for instance, if it is caged), the upper mandible grows so long it interferes with eating and other activities that are important to the chicken's well-being.

The upper mandible is naturally a little longer than the lower mandible. If the upper half begins to overlap the lower, trim it back with a fingernail file. Once it has passed the

filing stage, use toenail clippers or canine clippers to trim it back. Unless the upper beak has grown far too long, the part that needs to be trimmed away will be lighter in color than the rest of the beak. If in doubt, look inside the chicken's mouth to see where live tissue ends.

In most cases only the upper half needs trimming. Rarely, the lower part may need reshaping, especially if a too-long upper mandible has pushed the lower mandible in the opposite direction.

Beak trimming is not the same as debeaking, although it has been used euphemistically in the commercial poultry industry as a synonym.

overgrown upper mandible

correct

behavior modification \

Training a chicken to act in certain desirable ways or to do tricks, requiring a consistent, methodical approach and lots of patience. To modify a chicken's behavior, first determine exactly what you want the chicken to do, then develop a series of training sessions that gradually lead to your goal. Start with a simple step the bird can easily handle and in subsequent sessions gradually escalate toward the goal behavior. Keep training periods short (10 to 15 minutes) and consistent, and remain calm.

No books or DVDs specifically address training a chicken, but the general principles (known as operant conditioning) are the same as those used to train other animals. Do a keyword search for "training chickens" on the Internet, and you will find lots of information and some amusing video clips.

Belgian Bearded d'Anvers \

A breed of true bantam named after the Belgian municipality of Antwerp (which in French is called Anvers), where the breed was developed. The d'Anvers has a rose comb, small or no wattles, a full beard, and muffs that cover the earlobes. It comes in several color varieties. The hens lay small eggs with white shells and tend to go broody.

[Also called: Antwerp Belgian, Bearded d'Anvers]

Belgian Bearded d'Uccle \ A

breed of true bantam named after the Belgian town of Uccle, where the breed was developed, supposedly by combining Belgian Bearded d'Anvers and Booted Bantams to attain a breed that has both a beard and boots. The d'Uccle has a single comb, small or no wattles, a full beard and muffs, and fully feathered legs with vulture hocks. It comes in several color varieties. The hens lay a small whiteshell egg and readily brood.

Belgian Bearded d'Anvers rooster

Belgian Bearded d'Anvers hen

Belgian Bearded d'Uccle rooster

Best of Breed (BB, BOB) \ A show award given to the best specimen among the first-place winners in all the classes in a single breed, which would include cock, hen, cockerel, and pullet classes for each variety in that breed.

biddy \ A hen, generally one that is overly fussy.

bilateral gynandromorph \ A rare, naturally occurring chicken having both male and female characteristics and organs, such that one-half of the chicken's body looks like a cock and the other side looks like a hen. *[Also called: chicken rooster; half sider]*

bilateral tufted \ Having ear tufts on both sides of the head, as opposed to one side only, as occasionally occurs in Araucanas.

billing out \ Even though chickens have a beak, not a bill, the practice of using the beak to scoop feed out of a feeder onto the ground is often called billing out. *See also: beaking out*

biosecurity \ Disease-prevention management. Biosecurity technically refers to collective precautions taken to protect a flock from infectious diseases but is commonly used as a catchall word that includes all measures taken to protect chickens from harm of any sort.

bird flu \ See box on page 31.

Black Rock \ Trade name for black sex-link hybrid chickens.

black sex-link \ The offspring of a nonbarred cock with a barred hen, typically a red-based cock such as a New Hampshire or Rhode Island Red, and a Barred Plymouth Rock hen. At hatch all the chicks are black, but only a cockerel has a white spot on its head. The cockerels feather out to be barred like the dam, while the pullets remain solid black; both genders mature with some red feathers. Trade names include Black Star, Black Rock, and Red Rock (sometimes called Rock Red).

Black Star \ Trade name for black sex-link hybrid chickens.

Black Sumatra \ *See: Sumatra.*

blade \ The back section behind the last point of a cock's single comb. **See page 66 for illustration.**

blastoderm \ The beginning of embryo development, arising from a blastodisc that has been fertilized, at which time the cells organize into a set of clearly visible concentric rings.

blastodisc/blastodisk \ A pale irregularly shaped spot of cells on the surface of an egg yolk that, if fertilized with sperm, becomes a blastoderm. *[Also called: germinal disc/disk]*

bleaching \ The fading of color from the beak, shanks, and vent of a laying hen with yellow skin. The skin of a yellow-skin pullet contains a considerable amount of pigment, obtained from green feeds and yellow corn. Over time she uses this pigment to color the yolks of her eggs. As time goes by and the

B

bilateral gynandromorph

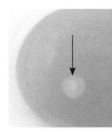

blastoderm

blastodisc

hen's old yellow tissue is replaced with new pale tissue, her skin, beak, and legs gradually become paler. After six months of intensive laying, the tissues of a high-producing yellow-skin hen will be nearly white.

bleaching sequence \ The order in which pigment fades from the body parts of a layer, based on how rapidly the tissue of each part is renewed. The more rapidly the tissue is replaced, the more quickly it turns pale. When a hen stops laying, color returns in the same order, approximately twice as fast as it disappeared. You can therefore estimate how long a yellow-skin hen has been laying, or how long ago she stopped laying, by the color of her beak, legs, and vent.

The first color loss is in the skin around the vent, which is renewed quite rapidly. Just a few days after a pullet starts to lay, her vent changes from yellow to pinkish, whitish, or bluish. Next to bleach is the eye ring. Within three weeks the earlobes of Mediterranean breeds bleach out; hens with red lobes do not lose ear color.

The beak's color fades from the corner outward toward the tip, with the lower beak fading more quickly than the upper beak. In breeds that typically have dark upper beaks, such as Rhode Island Red and New Hampshire, only the lower beak is a good indicator.

The best indicators of long-term production are the shanks, since they bleach last. Color loss starts at the bottoms of the feet, moves to the front of the shanks, and gradually works upward to the hocks.

BLEACHING SEQUENCE
(pigment returns in reverse order)

	Body Part	Eggs Laid	Average Time in Weeks*
1.	Vent	0–10	1–2
2.	Eye ring	8–12	2–2½
3.	Ear lobe	10–15	2½–3
4.	Beak	25–35	5–8
5.	Bottom of feet	50–60	8
6.	Front of shank	90–100	10
7.	Hock	120–140	16–24

From: *Storey's Guide to Raising Chickens,* 3rd edition

*Individual bleaching time depends on a hen's size, her state of health, her rate of production, and the amount of pigment in her rations.

BIRD FLU

The contagious viral infection bird flu has been around for many centuries. Over time numerous strains of the virus have evolved that are categorized as being of either low pathogenicity ("low path") or high pathogenicity ("high path"). Low-path strains generally cause minor or no signs in chickens and pose little health threat to humans. High-path strains spread more rapidly, are more likely to be fatal to chickens, and have the potential to affect humans. High-path bird flu has been detected in chickens — and eradicated — in the United States three times: in 1924, in 1983, and in 2004. No humans are known to have become ill in connection with these outbreaks.

A high-path virus is defined as one that kills 6 out of 10 chicks inoculated in a laboratory. The strain most often discussed in the news, H5N1, killed 10 out of 10 inoculated chicks. The letters H and N refer to two groups of proteins, the hemagglutinin, or H proteins, and neuraminidase, or N proteins. The 16 H proteins are identified as H1 through H16, and the 9 N proteins are N1 through N9. These two groups combine to form 144 different strains of the bird flu virus, each of which is identified by the two proteins they display. The H5NI strain has the fifth hemagglutinin and the first neuraminidase.

H5N1 does not normally affect humans, although since 1997, when it first appeared, sporadic human cases of a serious respiratory infection have occurred during bird flu outbreaks: People who became infected had extensive, direct contact with sick birds. Many people who did not get sick after contact with infected chickens developed antibodies to the virus.

A vaccine has been created to protect chickens against the known high-path strains, for use in the event of an outbreak to create a buffer zone around diseased flocks and prevent spreading of the flu while infected chickens are being destroyed. But like human flu viruses, bird flu viruses frequently mutate. So just as your annual flu shot doesn't always give you immunity against the flu strain that goes around, existing bird flu vaccines may not protect chickens against future bird flu strains.

The most common way bird flu spreads from chicken to chicken is through direct contact with droppings from infected birds and droppings on contaminated equipment and shoes. It is therefore more likely to occur in industrialized operations, where thousands of chickens are irresponsibly crowded into unsanitary environments, than in a small flock enjoying the benefits of sunshine and fresh air.

[Also called: avian influenza]

bleeding \ Allowing the blood to drain from a freshly killed chicken, to ensure that no trace of blood remains in the meat, for the best flavor, appearance, and keeping quality. Draining a bird's several tablespoons of blood must be done immediately after the kill, while the blood still flows freely. It is done either by chopping off the bird's head or slicing into its jugular vein with a sharp knife.

bleeding wound \ An injury such as a broken or ripped-off toenail or spur case can bleed profusely. Reduce or stop the bleeding with several puffs of a wound powder such as Wonder Dust. Flour or cornstarch is a handy substitute. As the bleeding lessens, use gentle pressure to hold a gauze pad to the wound until the blood flow stops, which shouldn't take more than about 10 minutes. Treat the wound with an antiseptic, and isolate the bird until it heals.

blinders \ A plastic device resembling a pair of tiny sunglasses, sometimes used to prevent or control cannibalism by blocking vision so a chicken can't see directly ahead to aim a peck. The use of blinders may lead to eye disorders.
[Also called: specs, peepers]

blinders

blood feathers \ *See: pinfeathers*

blood spot \ A small, dark, reddish or brownish dot that appears in less than 1 percent of all eggs. Each developing yolk in a hen's ovary is enclosed inside a sac containing blood vessels that supply yolk-building substances. A mature yolk is normally released into the oviduct from the only area of the yolk sac that is free of blood vessels, called the stigma or suture line. Occasionally, a yolk sac ruptures at some other point, causing vessels to break and a little bit of blood to get on the yolk. A blood spot may also be caused by a minor rupture in the wall of the oviduct, in which case it may appear anywhere within the white.

Spots sometimes occur in a pullet's first few eggs, while her reproductive system is still maturing, but are more likely to appear as a hen ages. Sometimes triggered by too little vitamin A in the diet, they are likely to be hereditary. A blood spot does not affect the egg's hatchability, but replacement pullets hatched from spotty eggs will most likely lay spotty eggs in turn.

A blood spot may look unappetizing, but it does not affect the safety of an egg for eating. As the egg ages, the spot becomes paler, so a bright spot indicates the egg is fresh. A blood spot is easily seen in an egg cracked into a bowl or frying pan, and it may be readily detected in an unbroken egg by candling.
See also: egg development

bloom \ The radiant glow of a breeder or exhibition chicken in peak condition. Constant showing causes stress, and stress is not conducive to bloom, so

experienced exhibitors who enter frequent competitions alternate or rotate the birds shown. \ The moist, light, protective coating on a freshly laid egg that dries so fast you rarely see it. Its purpose is to seal the pores in the shell to prevent bacteria from entering the egg and to minimize evaporation of moisture from inside the egg. Washing an egg removes the bloom and its protective qualities. Most of the pigment that gives the shell its color is in the bloom, which is why rubbing the shell of a freshly laid egg removes some of its color.
[Also called: cuticle]

blowout \ A serious condition in which a hen's uterine tissue remains prolapsed as a result of laying an oversize egg. If you act early, you may be able to reverse the situation by gently pushing the protruding tissue back inside, applying a hemorrhoidal cream such as Preparation H, and isolating the

hen until she heals. Left untreated, the exposed pink tissue will attract other birds to pick, and the hen will eventually die from hemorrhage and shock. Blowout that progresses to this stage is called pickout.

The situation may be largely prevented by avoiding breeds and hybrids selected for laying exceptionally large eggs, by proper feeding to prevent hens (especially heavy breeds) from growing fat, and by ensuring pullets don't start laying at too young an age. Pullets hatched from April through June mature at a normal rate, making them less likely to experience reproductive issues. Pullets raised out of season are at greater risk for early laying, which may be prevented through the use of controlled lighting. *See also: cannibalism; pickout; vent picking; controlled lighting*

blue \ A slate-gray plumage color found in two kinds of blue chickens. In

BREEDING STANDARD BLUES

Producing standard blue chickens involves a complex combination of genetic factors that result in splashed and black chicks, as well as blue ones.

Parents	Offspring		
	BLUE	**BLACK**	**SPLASHED**
BLUE × BLUE	50%	25%	25%
BLUE × BLACK	50%	50%	
BLUE × SPLASHED	50%		50%
BLACK × SPLASHED	100%		
BLACK × BLACK		100%	
SPLASHED × SPLASHED			100%

one kind — called lavender, self blue, or true blue — the birds are generally of uniform color and, when mated together, produce offspring of the same color. The more common kind of blue — called blue, Andalusian blue, or standard blue — results from a complex combination of genetic factors that modify black feathers so they appear to be blue laced in black, with the cocks darker than the hens. This color is characteristic of the Andalusian breed, as well as several varieties such as blue Cochin, blue Orpington, and blue Rosecomb. When two birds of this color are mated, only about 50 percent of the resulting offspring are blue.

Breeding standard blue chickens of the desirable color is difficult, as they can appear in many different shades, often with undesirable brassiness, mossiness, or smut. However, chickens may have both standard blue and self blue genes, making it possible to develop true blue chickens over several generations by selectively breeding birds having blue pigmentation in the down (as chicks), feathers, or legs and continuing to breed the best blue offspring. Breeding true blues carries its own difficulty, as it can eventually result in deformed wing and tail feathers.

blue Andalusian \ *See:* *Andalusian; blue*

blue egg \ An egg with a blue shell, such as is laid by an Ameraucana or Araucana hen. Since the blue-egg gene is dominant, a hen that lays blue eggs is not necessarily a purebred Ameraucana or Araucana. Similarly, an Ameraucana or Araucana hen that lays non-blue eggs is not purebred. For instance, mating a

blue-egg layer with a brown-egg layer results in hens that lay green eggs. A flock of Ameraucanas and Araucanas that lay eggs with green shells, or any color other than blue, may be brought back to laying blue eggs in a few generations by hatching eggs with the bluest shells, by using only breeder cocks that hatched from blue-shell eggs, and by breeding the offspring only to birds that lay eggs with blue shells.

See also: shell color

blue hen's chicken \ A strain of Old English Game birds that was famous during colonial days, when cock fighting was a national pastime. The offspring of a particular blue hen owned by Jonathan Caldwell of Kent County, Delaware, were known as especially fierce fighters. When the Revolutionary War started in 1775, Captain Caldwell and his Kent County company became known as Sons of the Blue Hen.

According to one legend, they acquired the nickname because they had taken along some of the blue hen's chickens to amuse themselves between battles. The story is pretty unlikely for two reasons: Fighting men on the move don't carry anything more than is absolutely necessary, and those men would have known that the cocks' inevitable crowing would give away their position to the enemy. A more likely version is that the company fought so valiantly they were nicknamed after their captain's famous fighting cocks.

During the Civil War another company from Kent County was called the Blue Hen's Chickens. In 1939 the blue hen chicken became the state bird, sponsored by the Delaware Chapter of the National Society of Daughters of Founders and Patriots of America.

Although no chicken today is a descendant of Caldwell's blue hen, over the years various fanciers within the state have developed strains of blue Old English in an attempt to replicate the legendary originals. Because the University of Delaware calls its various athletic teams the Fightin' Blue Hens, one such fancier felt the teams should have blue Old English as mascots and in the 1960s gave several pair to the university, where they are called Delaware Blue Hens. A professor there attempted to improve the color by crossing them with blue Andalusians. As a result today's Delaware Blue Hens are not the same as Caldwell's blue hen's chickens.

bluing \ A household product, sometimes sold as laundry blue or washing blue, used to brighten the color of white fabrics during laundering. Added to the final rinse when washing a chicken for show, bluing brightens the plumage of a white chicken. Use no more than two drops of bluing, or the bird may turn blue. To prevent streaking, thoroughly stir the bluing into the water before rinsing the bird.

body \ The remainder of a chicken after excluding the head, neck, wings, tail, thighs, legs, and feet. \ In describing a bird's shape, the body is considered to be the entire chicken. \ In describing a bird's plumage color, the body is considered to consist of the lower sides, abdomen, and stern but not the back and breast.

body temperature \ The normal body temperature of a chick is just under 107°F (41.7°C) and of a mature chicken is about 103°F (39.5°C). A chicken's temperature normally decreases slightly at night and during cool weather, and increases slightly during the day and in warm weather.

body type \ *See: type*

booster \ An additional dose of vaccine or toxoid periodically given to increase or renew the effectiveness of the first dose.

Booted Bantam \ A breed of true bantam that was developed in the Netherlands from bantams originating in Asia. Closely related to the Belgian Bearded d'Uccle, the Booted Bantam was once considered a nonbearded variety of the same breed. The breed comes in several color varieties, has a single comb, and has fully feathered legs with vulture hocks. The hens lay a small white-shell egg and readily brood.

Booted Bantam hen

boots \ Feathers on the shanks and toes, combined with vulture hocks, as is characteristic of the Belgian d'Uccle, Booted Bantam, and Sultan but undesirable in other breeds with feathered legs.

botulism \ A potentially fatal illness caused by toxins generated by *Clostridium botulinum* bacteria. These bacteria naturally live in soil and commonly occur in the intestines of chickens without causing disease, but they create powerful toxins as they rapidly multiply in the carcass of a dead bird or other animal or in a rotting cabbage or other solid vegetable. A chicken that gets poisoned from pecking at rotting organic matter or the maggots feeding on it or from drinking water containing rotting matter becomes gradually paralyzed from the feet up. It is also called limberneck because the progressive paralysis causes wings to droop and the neck to go limp. By the time the eyelids are paralyzed, the bird looks dead but continues to live until either its heart or respiratory system becomes paralyzed.

If the chicken isn't too far gone, it might be brought around with botulinum antitoxin, available from a veterinarian, or with a laxative flush to absorb the toxins and speed their journey through the intestines. Botulism may be prevented by promptly removing any dead bird or other animal found in the yard and by not feeding chickens rotting kitchen scraps or garden refuse.
[Also called: limberneck]

bow legs \ A deformity in which the legs, as viewed from the front or back, are farther apart at the hocks than at the feet.

boxing \ A bloodless variation of cockfighting in which the cocks' spurs are covered with foam rubber and the birds wear vests that electronically record hits, so they can spar without causing injury or being injured.

Brahma \ A breed developed in the United States from chickens that supposedly originated in India's Brahmaputra River valley. It is one of the largest breeds but also has a bantam version. The Brahma has a pea comb and feathered legs and comes in a few color varieties. The hens lay brown-shell eggs and brood easily.

brail \ To restrict flight temporarily by wrapping one wing with a soft cord or strap so it can't be opened for flight. Brailing is sometimes used to prevent a show chicken from flying so its

Brahma hen

appearance won't otherwise be marred by feather clipping or to ground a young bird that will eventually grow too heavy to fly. The cord must be moved to the opposite wing every two weeks so it doesn't get so tight it causes damage and so the bird can exercise its wing muscles. \ The cord used for brailing.

brassy \ An undesirable yellowish metallic hue in red, white, or blue plumage and less commonly in other colors, especially in a cock's hackle, wing bow, and saddle. Brassiness may be hereditary or the result of diet but is most often induced by excessive exposure to direct sunlight.

break up \ To discourage a hen from brooding so she will continue to lay eggs. Depending on how determined the hen is to set, she may be broken up with one or more of the following measures:

- Collect eggs often so they don't accumulate in the nest.
- Repeatedly remove the hen from the nest.
- Move or cover the nest so she can't get to it.
- House the hen in a different place.
- Confine the hen to a broody coop.

breast \ The front underside of a chicken's body from the base of the neck to between the legs, including the muscle on both sides of the breastbone.

breast blister \ A blister resulting from pressure on the breastbone's keel from the ground or other hard surface, commonly seen in heavy-breed males and in broilers. The blister is uncomfortable for the bird but does not pose a serious

health risk unless it becomes infected. A blister does, however, mar the appearance of a broiler raised for meat.

Breast blisters may be prevented by housing heavy breeds on soft litter or grassy pasture rather than on wire or hard-packed bedding and by removing roosts. Breast blisters may be minimized in broilers by reducing their growth rate. [*Also called: keel bursitis; keel cyst; sternal bursitis*]

breastbone \ The large, bony plate covering more than half of the front part of a chicken's body to protect its internal organs and provide a place for attachment of the flight muscles. A chicken's breastbone is long in comparison to swimming birds, which have a wide breastbone, and flying birds, in which the breastbone is nearly as wide as it is long. The end of the breastbone is cartilage in a young chicken and hardens as the bird matures, providing a way to determine if a particular bird is young or old.

See pages 18 (anatomy) and 44 (butchering) for illustrations.

[*Also called: keel bone; sternum*]

breed \ To mate a cock and hen (or hens), for the purpose of obtaining fertile eggs. \ A genetically pure line having a common origin, similar weight, carriage, conformation, and other identifying characteristics, and the ability to reliably produce offspring with the same characteristics. Most chicken breeds fall into one of three categories:

STANDARDIZED BREEDS are those for which a formal description has been published. They generally have limited genetic diversity and therefore limited variability, although many standardized breeds are subdivided into two or more distinct varieties. Standardized chickens are bred to conform in appearance to a formal description for their breed, which may vary from country to country. A standardized breed may be deliberately created through selective breeding or be formalized from a landrace breed. Nearly all chicken breeds in North America are standardized.

LANDRACE BREEDS are isolated local populations that have developed over many centuries. They have greater genetic diversity than standardized breeds and much more variability. When deliberate selective breeding is involved, these chickens are bred to produce well in a specific natural environment. The few landrace chicken breeds in North America are recent imports.

INDUSTRIAL BREEDS are those chosen for the mass production of meat or eggs. They were originally developed from standardized varieties, largely, white varieties of Cornish, Leghorn, and Plymouth Rock, but have a more limited gene pool — by some accounts lacking more than half the genetic diversity available to the species — and even greater uniformity. Industrial breeds are bred for consistency in laying ability or growth rate and for efficient feed conversion in a controlled environment.

breed club \ An organization devoted to promoting a specific breed on a national or regional scale. Most groups are run by volunteers and therefore come and go according to the enthusiasm of the current crop of staffers. The best way to locate a currently active breed club is to search the Internet.

breeder \ A mature chicken involved in the production of fertile eggs for the purpose of hatching the next generation. \ A person who manages such chickens.

breeder ration \ A complete feed designed to fulfill all nutritional requirements for the production of eggs for hatching. *See also: feeding breeders*

breeder quality (BQ) \ A chicken with minor defects that would bar it from winning at a show, but with the genetic potential to produce show-quality offspring.

breeding quartet \ One cock mated with three hens of the same breed and variety. *[Also called: quartet]*

breed true \ To produce offspring that are exactly like the parents.

breed type \ *See: type*

bridge \ Feed that has packed tight and remains suspended in a tube feeder, creating an empty area at the bottom after the feed below has been eaten. Thus, feed remains in the feeder but the chickens have nothing to eat. Compared to crumbles or pellets, mash tends to pack and bridge in a tube feeder that's filled to more than two-thirds capacity.

broiler \ A young, tender chicken raised for meat. Commercial-strain broilers are ready to butcher at 5 to 6 weeks of

age; utility breeds may take 12 weeks or more. Management issues pertaining to homegrown meat birds primarily relate to rapid growth and heavy weight. Common issues for commercial-strain broilers are lameness, breast blister, and heart failure; for utility breeds, breast blister is the most common concern.
[Also called: fryer; meat bird]

broiler ascites \ *See: ascites*

broiler house \ A building for confining a large number of chickens raised for meat.

bronze \ An undesirable red-brown metallic hue in solid red, black, or white plumage.

brood \ To sit on eggs until they hatch. \ To care for newly hatched chicks. \ The chicks themselves.

brooder \ A mechanical device that supplies chicks with warmth and protection similar to what a mother hen provides. A brooder may be homemade or ready-built, small or large, stationary or portable, and heated by electricity, oil, hot water, or hot air. All the various styles fall into three basic categories:

BOX BROODER. A single, portable unit that is generally square or rectangular (hence the name) but may be round or oval. A ready-built box brooder is made of metal and/or plastic; a homemade brooder may be fashioned from a container of suitable size, such as a sturdy cardboard box or a galvanized livestock water tank with a lightbulb or heat lamp furnishing warmth.

BATTERY BROODER. A ready-built unit consisting of a series of box brooders stacked one on top of the other, allowing a large number of chicks to be brooded in a limited space. The individual units may used to house chicks of different ages or different breeds.

AREA BROODER. A portion of a barn, shed, or other outbuilding with a heat source surrounded by a brooder guard. An area brooder is an economical way to raise large numbers of chicks and allows them to be started in the same facility where they will live at maturity.

brooder guard \ A 12- to 18-inch (30 to 45.7 cm)-high corrugated cardboard fence that confines brooded chicks to within 2 to 3 feet (0.6 to 0.9 m) of a heat source.
[Also called: chick corral]

brooder pneumonia \ Aspergillosis.

brood

BROODING

Successfully rearing baby chicks in the same manner as the mother hen requires:

ADEQUATE SPACE. Start with about 6 square inches (40 sq cm) per chick, and as they grow increase available space for sanitary reasons and to prevent boredom that leads to picking. Rate of growth varies from breed to breed, so adjust living space accordingly.

RELIABLE HEAT. Start at approximately 95°F (35°C) and reduce the temperature approximately 5°F (3°C) each week until it is the same as the ambient temperature. When heat is furnished by incandescent lights, start with 60- or 100-watt bulbs and adjust the temperature by raising or lowering the fixture or by decreasing or increasing the wattage. When heat is furnished by an infrared lamp, start the lamp about 18 inches (45 cm) above the chicks, and raise it about 3 inches (7.5 cm) each week. If the heater is a hover or heating panel, start it about 4 to 6 inches (10–15 cm) above the floor, or just high enough for the chicks to walk under without bumping their heads, and raise it about 3 inches (7.5 cm) each week. The temperature may be measured with a thermometer 2 inches (5 cm) above the brooder floor or by monitoring chick activity. If they crowd together near the heat and peep shrilly, they're cold; if they crowd together away from the heat and pant, they're too hot; if they peep musically while moving around to eat and drink and sleep sprawled side by side, the temperature is just right.

APPROPRIATE LIGHT. To help chicks find feed and water, furnish continuous light for the first 48 hours. Thereafter, if the brooding area gets natural daylight through a window, the light may be turned off during the day unless the source of light is also the source of heat. In that case turn it off for half an hour during each 24-hour period — but obviously not during the coolest hours — so the chicks learn not to panic and pile up later when the lights are turned off at night or in the event of a power failure.

FEED AND WATER. Chicks can go without water for 48 hours after hatch, but the sooner they drink, the better they will grow. Likewise, they can survive a couple of days without eating but should eat soon after taking the first drink. Feeders and drinkers must be of a size and style designed for chicks, of sufficient number for the quantity of chicks being brooded, and positioned where the chicks can readily find them. *See also: feeder, chick*

A CLEAN ENVIRONMENT. As well as insulating the floor for added warmth, bedding absorbs moisture and droppings. Ideal kinds of bedding are peat moss, pine shavings, chopped straw, finely shredded paper, pellet bedding, vermiculite, and coarse sand. For the first two days, until the chicks learn to eat starter and not bits of bedding, use paper towels. To make the surface firm, put down a few layers of newspaper or opened-out feed sacks, then keep the chicks from slipping on the slick paper with a layer of paper towels. Add a fresh toweling as often as necessary to maintain cleanliness, and after two days roll up all the paper for disposal along with loose bedding. Thereafter, to maintain clean bedding, stir bedding often so it won't pack, and add fresh bedding as often as necessary to keep it fluffy and absorbent.

PROTECTION FROM MOISTURE. Damp bedding gets moldy and can cause brooder pneumonia. Remove and replace moist bedding around waterers, and ensure they are level to prevent leakage. Damp chicks are vulnerable to becoming chilled. Make sure chicks have sufficient space to spread out for sleeping so none get crowded into the drinker.

GOOD VENTILATION. Chicks generate a lot of moisture through respiration and droppings. Good ventilation allows moisture to evaporate and also brings in fresh oxygen to keep chicks healthy. Homemade brooders, particularly those fashioned from plastic storage totes, are especially likely to lack adequate ventilation.

INSULATION FROM DRAFTS. Chick down provides some measure of insulation to help chicks retain heat, but air movement can remove the trapped heat. If air movement can be detected at chick level, the chicks most likely feel a chilling draft. Move the brooder away from open windows and other sources of draft; lay a piece of cardboard partially across the top of a homemade brooder; adjust the brooder guard in an area brooder.

PROTECTION FROM PREDATORS. Chicks are especially vulnerable to predators that can sneak through cracks and other small spaces, such as rats and snakes. Raccoons can pry open doors that aren't latched shut. And family pets should never have unsupervised access to brooding chicks.

FREEDOM FROM STRESS. Chicks' immunity is decreased by stress, which both reduces their growth rate and makes them susceptible to diseases they might otherwise resist. In addition to meeting all the above requirements for successful brooding, other stress reduction measures include not handling chicks for their first few days of life, talking or humming whenever you approach so the chicks won't be startled by your sudden appearance, and leaving the old feeders and drinkers for a few days whenever they are being replaced by larger ones as the chicks grow.

Shown here in a cutaway, chicks may be brooded in a sturdy cardboard box with a light bulb for warmth.

brooding nest \ An isolated place where a setting hen can brood without danger of another hen taking over her nest when she gets off to eat and without the danger that other chickens might kill the newly hatched chicks, much as they would destroy a mouse or frog found in the yard. The brooding nest should be darkened; well ventilated; and protected from wind, rain, and temperature extremes. It should be 14 inches (36 cm) square and at least 16 inches (41 cm) high, with a 4- to 6-inch (10 to 15 cm)-high sill at the front to hold in nesting material. Clean, dry wood shavings, except cedar (which is toxic), make the best nesting material. Be sure both litter and hen are free of lice and mites, which can make a hen restless enough to leave the nest. Once the hen has shown she is determined to remain broody, move her after dark, and she'll be less likely to try to get back to her old nest.

If you have more than one hen brooding at a time, move them to separate brooding nests; otherwise they may both pile into one nest and abandon the other. Or they may accidentally switch nests, shortening the incubation period for one and prolonging it for the other, and the latter hen may get discouraged and quit before the chicks hatch. Or they may both follow the first chicks that hatch, leaving the remaining eggs to chill and die. Avoid these problems by separating broody hens.

brooding pen \ Scaled-down chicken housing where a hen may safely brood her chicks away from other chickens that might pick on the little ones; where they are well protected from house cats, hawks, and other predators; and where they can neither wander away and get lost nor become chilled in tall, dew-dampened grass. Feed as you would for baby chicks. The starter won't hurt the hen, but layer ration with its high calcium content would damage the chick's tender kidneys. Use a drinker the hen can't knock over. A 1-gallon (3.75 L) waterer is ideal. When brooding bantams, put marbles or pieces of clean gravel in the rim for the first few days to prevent a tiny chick from drowning.

brood patch \ A large defeathered bare area on a setting hen's breast. Its purposes are to bring her body warmth closer to the eggs and to keep the eggs from drying out too fast by lending moisture from her body.

broody \ Having the tendency to incubate a clutch of eggs until they hatch. The instinct to brood is generally triggered by decreasing day length and encouraged by seeing an accumulation of eggs in a nest. Signs of broodiness include clucking like a mother hen with chicks and puffing out the feathers and hissing or growling when disturbed on the nest.

When a hen gets broody, her pituitary gland releases the hormone prolactin, which causes her to stop laying. Light, flighty breeds selectively bred for high egg production are therefore less likely to brood than other breeds; individual hens may occasionally brood but are not always reliable. Heavy breeds make good broody hens that can handle large numbers of eggs, although a really heavy hen with a loaded nest may break some of the eggs. Most hens can cover

12 to 18 eggs of the size they lay. A bantam can hatch only 8 to 10 eggs laid by a large hen, while a large hen might cover as many as two dozen bantam eggs.

broody coop \ A swinging cage, with a wire or slat floor, where a broody hen is housed for as long as it takes for her to break up — usually 1 to 3 days. The longer the hen has been broody, the longer she'll take to start laying again. A hen that's broken up after the first day of brooding should begin laying in 7 days; a hen that isn't broken up until the fourth day may not lay again for 18 days or so.   *See also: break up*

broody growl \ A low, intense, harsh defensive sound made by a setting hen that's been disturbed on the nest. It is accompanied by feather ruffling to increase the intimidation factor and may include a peck — for instance, to a human hand reaching under the hen to retrieve an egg. The same sound is sometimes made by a hen with chicks in protest to a cock intent on mating; a low-ranking hen approached by a higher-ranking hen; or any hen on seeing a small, familiar animal such as a cat or rat.

broody hiss \ The hissing sound made by a setting hen that's been disturbed on the nest. Similar to a snake's hiss, the sound indicates the hen is wary and has her guard up.

broody poop \ A big blob of manure dropped by a setting hen when she steps off the nest. To avoid contaminating the eggs, a hen rarely poops in the nest but holds it until she leaves the nest seeking water and nourishment.

brown egg \ *See: shell color*

Buckeye \ A breed developed in the United States, in the Buckeye State of Ohio, as a dual-purpose farmstead chicken that adapts well to cold weather. The Buckeye has a pea comb and comes in both large and bantam sizes in a single color — a rich reddish brown of about the same shade as a buckeye nut. The hens lay eggs with brown shells and brood easily.

buffalo wings \ Fried chicken wings served as an appetizer with hot sauce, along with celery sticks and blue cheese dressing to reduce the spiciness. The dish is named after Buffalo, New York, where it originated.

Buckeye rooster

built-up litter \ See: composting litter.

bumblefoot   See: foot pad dermatitis

bursa of Fabricius \ The cloacal bursa, named after the anatomist Hieronymus Fabricius who first described the organ in 1621.

butt-end band \ A leg band, called butt-end because the two ends butt together instead of overlapping. Sealing them requires a different applicator tool for each size. These bands are designed as a permanent means of identification and may be applied only to chickens reaching full size. They are the most expensive leg bands and the most elegant looking.

Buttercup \ See: Sicilian Buttercup

buttercup comb \ The cup-shaped comb characteristic of the Sicilian Buttercup. It consists of a single blade at the front, above the beak, that divides to form a closed crown centered on top of the skull, with a circle of evenly spaced points around the outside rim but none arising from the center. It somewhat resembles a teacup with serrated edges. [*Also called: cup comb*]
See page 67 for illustration.

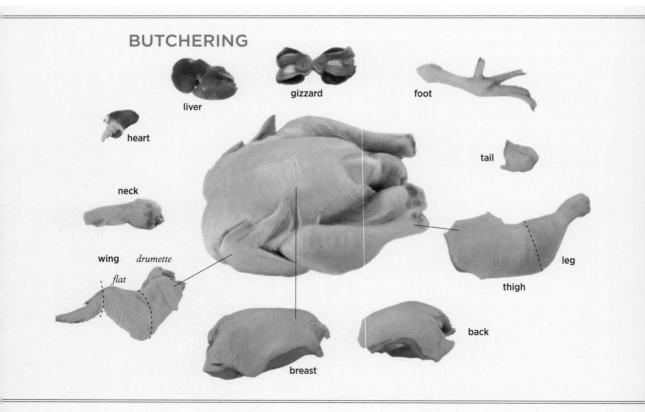

BUTCHERING

liver

gizzard

foot

heart

tail

neck

wing   *drumette*

flat

breast

back

thigh

leg

cackle \ A series of short, sharp sounds made by both cocks and hens as an alarm when something unusual is spotted and as a warning of potential danger. The most intense cackling is done by a hen upon leaving the nest after laying an egg, perhaps to scare away any predator that may be lurking, as well as to put other chickens on notice that she may need help should a predator in fact have sneaked up while she was occupied in the nest.

cackleberry \ A humorous alternative word for egg. It is called a cackleberry because of the characteristic cackling made by some hens upon leaving the nest after laying an egg.

cage confinement \ Housing chickens within a cage or hutch with no access to an outside yard or run. Chickens may be caged for any number of reasons. Commercial laying hens are caged to control their diet and guard against diseases, predators, and weather extremes. Exhibition birds are caged while being trained and conditioned for the showroom and to control breeding. Breeder cocks may be caged to keep them from fighting with one another or defeathering hens; in such a case the cocks might be rotated so each has a turn running with the hens. A defeathered hen might be caged until her feathers grow back. A sick or injured chicken might be caged until it recovers. A mother hen might be penned up with her chicks to protect them from cats and other predators. Urban and suburban pet chickens might be caged to keep them from annoying the neighbors or being eaten by a dog, although in this case the chickens are often let out under the owner's supervision.

The cage should be big enough that the bird can move around freely without rubbing against the sides or top. Heavy breeds should never be housed in an all-wire cage, and no chicken should be kept on suspended wire as permanent full-time housing. If a mature chicken needs to be caged, make the cage floor of wooden batten boards or of solid plywood covered with bedding, or at least provide a resting pad such as those used for rabbits.

cage confinement

calcium grit

cage free \ Not confined to a cage but likely confined within loose housing.

calcium carbonate \ A calcium supplement derived from such sources as oyster shells, aragonite, and agricultural limestone — but not dolomitic limestone, which can be detrimental to egg production.

calcium grit \ A source of calcium carbonate, such as crushed oyster shell, that erodes more readily than an inert grit such as crushed granite, but less readily than other forms of calcium carbonate such as aragonite, and therefore serves double duty as both a source of calcium and a source of grit for the gizzard. Calcium grit is sold commercially, sometimes in a variety of screened sizes, one of which is called chick grit or starter

CALCIUM

Calcium is an essential constituent of bones and eggshells. A laying hen uses some calcium from her diet to develop an eggshell and draws the rest from her own bones.

The amount of dietary calcium a hen needs varies with her age, diet, and state of health. Older hens need more calcium than younger hens because their bones have been depleted. Hens on pasture obtain some calcium by eating beetles and other hard-shelled bugs but may not obtain enough to meet their needs.

In warm weather, when chickens eat less, the calcium in their regular ration may not be enough, resulting in thin-shelled eggs. Hens that eat extra ration in an attempt to obtain sufficient calcium get fat and become poor layers. But calcium should be offered separately and free choice, not mixed into the ration, to allow for differences in the needs of individual hens, because those chickens that need less calcium may overdose, resulting in kidney damage.

Thin-shelled eggs or brittle and broken feathers indicate a calcium deficiency, which can occur even if a calcium supplement is offered. To absorb the calcium hens also need dietary phosphorus, and to metabolize both they need vitamin D, which they receive from just half an hour a day of direct sunlight.

Eggshells contain exactly the right balance and may be recycled back to the hens to help replenish lost calcium and other nutrients. Wash the shells, dry them, and crush them before feeding them to hens. Feed only dry shells; shells that remain moist may harbor harmful molds and bacteria. Shells may be dried by spreading them on a sheet of paper in the sun, stirring occasionally, for two or three days; by spreading them on a baking sheet and putting them in a 300°F (150°C) oven for 10 minutes; or by microwaving them on high for a minute and a half.

grit. But calcium should never be fed to birds that are not nearing age of lay, as it can interfere with bone development, as well as damage the liver. Similarly, any chicken (such as a cock or a poor laying hen) that does not need a calcium supplement can overdose on calcium by eating calcium grit because they have no access to inert grit.

[Also called: mineral grit]

California Gray \ A breed developed in the United States by crossing the white Leghorn and the Barred Plymouth Rock to attain a dual-purpose chicken that is heavier than a Leghorn and lays nearly as well but is not as flighty. It has no bantam counterpart. The California Gray has a single comb and comes only in a black-and-white cuckoo color pattern that is autosexing — the pullets are darker than the cockerels; at maturity, the hens' barring is white and dark gray, the cocks' barring is white and light gray. The hens lay white-shell eggs and seldom brood. California Grays are used primarily to produce California White hybrid layers.

[Also called: production black]

California White \ The only sex-link hybrid that lays white-shell eggs, produced by crossing a California Gray cock and a White Leghorn hen. The result is a white chicken with a few black feathers that can be sexed at hatch. All the chicks are yellow with black body spots, but only the pullets have head spots. The hens lay almost as well as Leghorns but adapt better to cold climates.

calls \ The various sounds chickens make. Sounds made by chicks and by cocks have been studied more than those made by hens. To date, some 24 to 30 distinct calls have been identified by various researchers, although no one has developed a definitive list or determined with certainty the function of each call.

C

California Gray hen

Campine \ A breed developed in the northeastern part of Belgium known as Campine. It has tight feathering and a single comb, and the hen's comb flops to one side. Both the large and bantam versions come in two color varieties: golden (rich golden head and neck feathers, with black and gold barring in the body feathers) and silver (the same pattern in white instead of gold). The cocks have a modified form of hen feathering: Their color pattern is identical to that of the hens, and their sex feathers are not as long and pointed as those of other breeds. The hens are good layers of white-shell eggs and seldom brood.

Campylobacter \ One of several species of bacteria of the genus *Campylobacter*. The species most likely to affect chickens, as well as humans, is *C. jejuni*. These bacteria colonize the intestines of chickens and other poultry and are spread through infectious droppings in feed and water. Although some strains of *C. jejuni* cause enteritis and death in newly hatched chicks, the bacterium is generally nonpathogenic in mature poultry.

campylobacteriosis \ An intestinal disease, caused by *Campylobacter*, that infects a wide range of animals, including humans and chickens. *C. jejuni,* in particular, causes this diarrheal illness in humans who eat undercooked infected poultry, primarily from commercial broiler flocks.

Campine hen

Campine rooster

candle \ To shine a light into an intact egg to examine its contents through the shell and determine its interior quality (for eating) or embryo development (for hatching); originally done with a candle. Poultry-supply outlets offer a handheld light designed for the purpose, but a small, bright flashlight works just as well.

White shells are easier to see through than colored shells, and plain shells are easier to see through than speckled shells. In a darkened room, grasp an egg at the smaller end between your thumb and first two fingers. With the egg at a slant, hold the large end to the light.

When candling a fresh egg for eating, turn your wrist to give the egg a quick twist, sending the contents spinning. The albumen of a fresh egg thins as the egg ages. If the yolk looks vague and fuzzy, the thick white albumen surrounding it is holding it properly centered within the shell. If the yolk is clearly visible, the albumen has thinned, allowing the yolk to move closer to the shell.

Candling to measure the air cell also gives you an idea of an egg's freshness. The cell of a freshly laid cool egg is no more than ⅛ inch (0.3 cm) deep. Above that, the larger the cell, the older the egg. Just how fast the air space grows depends on the porosity of the shell and on the egg's storage temperature and humidity. *See also: air cell*

Candling is also used to monitor the progress of hatching eggs, so those that are spoiled or nondeveloping may be removed from the incubator. *See also: embryo*

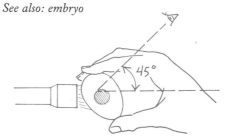

CANDLING INCUBATED EGGS

After one week of incubation, candling reveals one of three things:

- Nothing (or a vague yolk shadow) — either the egg is infertile or the embryo has died   *See: infertility*
- A thin ring within the egg or around the short circumference — the embryo has died   *See: fertility, weak*
- A web of vessels surrounding a dark spot — the embryo is developing properly   *See: embryo*

After two weeks of incubation (removing any infertile eggs or ones with blood rings), candling reveals one of two things:

- Murky or muddled contents that move freely and/or a jagged-edged air cell — the embryo has died   *See: embryo death*
- A dark shadow except in the air cell at the large end (perhaps movement against the air-cell membrane) — the embryo is developing properly

candler \ A light used for candling eggs. \ A person who candles eggs.

canker \ A disease caused by *Trichomonas gallinae*, protozoan parasites that primarily affect the mouth, throat, and crop. It is characterized by cheesy accumulations in the throat, usually accompanied by weight loss. It is more prevalent among domestic pigeons but spreads to chickens by means of stagnant drinking water or feed contaminated with discharge from an infected bird's mouth. Prevention involves good sanitation and avoiding birds that may be carriers. This illness does not affect humans and is not the same as trichomoniasis in humans.
[Also called: roup; trichomoniasis]

cankers \ Soft, whitish, cheeselike bumps that erupt to form sores, usually on the face or in the mouth.

cannibalism \ The nasty habit chickens have of eating each other's flesh, feathers, or eggs. Most cannibalism occurs at one of two specific periods in a chicken's life — as a chick when it gets its first feathers and as a pullet when it is beginning to lay. Cannibalistic picking is different from peck-order fighting to establish dominance, although frequent fighting to adjust the pecking order may lead to bloody injuries, which can in turn lead to picking and cannibalism.

Stressful situations that can cause irritability and cannibalistic picking include overcrowding, boredom, lack of exercise, bright lights, excessive heat without proper ventilation, too little perching space, too few feeding or watering stations, feeders or drinkers being placed too close together, nutritional imbalance, external parasites, and insufficient opportunities to engage in the natural activities of scratching and pecking the ground.

Once picking starts, it can be difficult to control. First, separate chickens that have been picked to the point of having bare patches or bloody wounds, since these picked areas invite more picking. Then identify and remove the instigators.

FORMS OF CANNIBALISM

Form	Likely Group
Toe picking	Chicks
Tail pulling	Growing birds
Feather picking	Growing birds
Vent picking	Pullets
Head picking	Cocks; any birds in adjoining cages
Egg eating	Hens

Switch to red lights that make blood more difficult to see, or change bright lights to dim lights. In hot weather open windows or turn on a fan to stir the air and reduce the temperature.

Since salt deficiency can cause chickens to crave blood and feathers, add 1 tablespoon of salt per gallon of water in the drinker for one morning, then repeat the salt treatment three days later. At all other times provide plenty of fresh, unsalted water.

Young birds raised on wire are more likely to become cannibalistic than those raised on litter, perhaps picking at each other as a substitute for pecking the ground. The most common form of cannibalism — picking at newly emerging blood-filled feathers, particularly on the back near the tails of chicks growing their first set of feathers — may be controlled by applying an antipicking cream, a diaper rash cream such as Desitin, or pine tar until the feathers grow in. *See also: blowout; pickout; vent picking*

cape \ The short, narrow feathers between a chicken's shoulders, where the neck joins the back, and growing underneath the hackle. As a group, these feathers grow in the shape of a cape.

capillariasis \ Capillary worm infection.

capillary worm \ A hairlike roundworm (nematode) that invades the crop and upper intestine, causing droopiness, weight loss, diarrhea, and sometimes death. When chickens sit around with their heads drawn in, capillary worms are the likely culprits.

capon \ A cockerel raised for meat that has been castrated so it will grow larger, plumper, and more tender. Compared to an intact male, a capon rarely crows and has an undeveloped comb and wattles; longer hackle, saddle, and tail feathers; and a calmer disposition. A capon is not properly finished until at least 20 weeks of age and, because of its excessive weight, is susceptible to weak legs and breast blister. For these reasons, and because hybrid broilers grow out quicker and are therefore cheaper to raise for meat, and because castration is now considered inhumane, capons are not as common as they once were.

rooster

capon

caponize \ To castrate a rooster.

carbohydrate \ Any of a large group of organic compounds occurring in feedstuffs that release energy when broken down through digestion. In most chicken rations, starchy grains such as corn, oats, and wheat are the primary source of carbohydrates.

carnation comb \ A comb that looks like a single comb at the front but thickens toward the back and sprouts random protruding spikes, giving it the appearance of a blossom or crown. This comb style is unique to the Empordanesa and Penedesenca breeds originating in Spain's Catalan region.
[Also called: clavel/clavell comb; crown comb; king's comb; king's crest]

carnation comb

C

Catalana hen

carriage \ A chicken's posture, which is characteristic of its breed and may tend toward the horizontal (for example, a New Hampshire or Welsumer) or the vertical (a Malay or Shamo). \ The angle of a chicken's body part, such as its back or tail, with respect to horizontal.

carrier \ A safe container for transporting chickens. \ An apparently healthy chicken that transmits disease to other chickens.

Catalana \ A breed developed in the Principality of Catalonia in Spain as a dual-purpose chicken that does well in hot weather. The Catalana has a single comb, and the hen's comb flops to one side. Both the large and bantam versions come in only one color — golden buff with a black tail. The hens lay eggs with white or slightly tinted shells and seldom brood.

CARRYING A CHICKEN

To hold or carry a calm chicken, reach beneath the bird and firmly grasp both its legs with one hand (so it can't scratch you), then bring your other arm around to cradle it. Once you have the chicken safely cradled, reassure it by stroking its wattles and throat.

To hold or carry a chicken that struggles to get free, firmly grasp both legs with one hand and rest its weight on the same arm, then with the other hand hold both wings over its back. Do not carry a chicken by its wings alone, which may cause damage to the bird or cause it to injure you with flailing claws and spurs. If the bird doesn't stop struggling when you stroke its throat and wattles, cover its eyes with a handkerchief laid over its head to help it calm down.

carrying a frightened chicken

carrying a tame chicken

catch a chicken \ A chicken that's
used to being around humans is easy to
catch. Just bend down and pick it up.
A chicken that isn't tame may easily be
lifted off its perch at night. During the
day catching an untame chicken is easiest with the aid of a catch net, catching
hook, catching crate, or at least a helper
to assist in cornering the bird. When you
and your helper get close enough, one of
you should block the bird's escape while
the other clamps down on its two wings.
Once you start trying to catch an untame
chicken, don't give up or you will have
a more difficult time catching it in the
future.

catching crate \ A lightweight
but sturdy wooden box or wire cage with
an opening at one end that fits snugly
against the outside of the pop hole and a
slide-down door that is closed after one
or more birds are herded from the coop
into the crate. A catching crate comes in
handy when you want to avoid exciting
or injuring untame chickens that need to
be caught. Some catching crates have a
second door at the top for removing the
bird(s). The crate may be used in a pen
or pasture with a pair of panels to guide
the chickens toward the crate.

catching hook \ A hook used to
nab a chicken on the run. It consists of
a plastic or wooden handle attached to
heavy-gauge wire bent into a hook. Snare
the chicken by one leg, pull it toward
you, and quickly pick it up while it's off
balance. Using a catching hook requires
quick thinking and physical agility.
[Also called: fowl catcher]

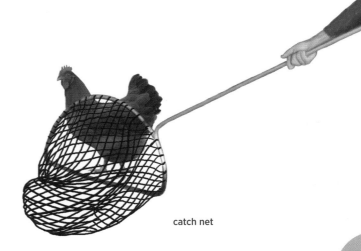

catch net

catch net \ A net for catching chick-
ens on the ground or on the fly, consisting of a netting pocket attached to a
metal hoop with a handle. Like a but-
terfly net or a fish net, the traditional
poultry net is framed by a circular hoop,
but unlike netting butterflies in the
air or fish in water, chickens are most
often netted on the ground, making a
D-shaped frame easier to use.

When you get close enough to
the chicken, scoop it into the net, then
quickly retrieve the bird. A really wild
chicken might slip back out, which can
be avoided by clamping the frame against
the ground and holding it down until
you are in a position to get the chicken
out of the net. Some athletic and well-
coordinated souls can catch a chicken in
midair and quickly retrieve it from the
net before it has a chance to escape.

cavernous nostrils \ The promi-
nent and deeply hollowed nostrils typical of a breed sporting a duplex comb
(except the Buttercup), compared to the
slit-shaped nostrils of other breeds.

cecal dropping \ A pasty yellowish, greenish, or brown dropping that occurs two or three times a day, or approximately every tenth dropping, as a result of the ceca emptying their contents. A cecal dropping is readily distinguished from the more frequent intestinal droppings because it is looser and smellier and lacks the white cap. The typical texture, color, and frequency of cecal droppings indicate normal digestive function.

cecal worms \ Short parasitic roundworms (nematodes) that invade the ceca, in a severe case resulting in weight loss and weakness. Although cecal worms are the most common nematode, they rarely cause a serious problem, and typically, a chicken carrying them exhibits no signs.

cecum (plural: ceca) \ One of two finger-shaped pouches at the juncture of the small and large intestines that resemble the human appendix. These blind pouches get their name from the Latin *caecus*, meaning blind. In the ceca, moisture is absorbed from digesting matter, and fermentation breaks down coarser cellulose.

cell autonomous sex identity \ Inherent molecular differences between the cells of a cock and a hen that control the development of sexual traits during embryonic growth, as distinct from sexual traits that are determined by hormones after the embryonic formation of male (testis) or female (ovary) sex organs.

cervical dislocation \ Killing a chicken by twisting, crushing, or stretching its neck.
[Also called: neck wringing]

cestode \ *See: tapeworm*

cestodiasis \ Tapeworm infection.

chalaza (plural: chalazae) \ One of two cords anchoring the chalaziferous layer of albumen to the shell membrane within an egg, thus protecting the yolk by centering it within the white. The cords are formed as the egg travels through the oviduct and rotates, twisting the ends of the chalaziferous layer (or inner thick) on opposite sides of the yolk. When an egg is broken into a dish, the chalazae snap away from the shell membrane and recoil against the yolk like two little white knots, giving these cords their name from the Greek word *khalaza*, meaning small knot.

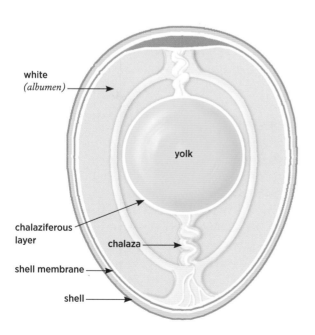

white (albumen)

yolk

chalaziferous layer

chalaza

shell membrane

shell

chalaziferous layer \ *See: inner thick*

Chantecler \ The only chicken developed in Canada; a dual-purpose breed that can withstand the rigors of the country's winters. The conformation of the Chantecler is strikingly similar to that of the Buckeye, developed in the United States with the same goal, except the latter has a pea comb. The name Chantecler is derived from the French words *chanter*, meaning to sing, and *clair*, meaning clear, and comes from the play *Chantecler*, in which the lead character is a rooster who brags that his clear song brings the sun up each morning.

The Chantecler was developed by judiciously crossing several breeds to obtain a white chicken with tight feathering that lays well in winter and has a frost-resistant cushion comb and small wattles. Later a partridge variety was developed to be more predator resistant on pasture. Both varieties come in large and bantam sizes. The hens lay brown-shell eggs and tend toward broodiness as they age.

check \ An egg with a cracked shell but with the shell membrane still intact. A check egg is safe to eat, provided it is refrigerated promptly and used right away.

chick \ A baby chicken until it begins to feather out.

chick corral \ *See: brooder guard*

chicken \ A male or female domesticated bird descended from jungle fowl and developed in numerous breeds for its

Chantecler rooster

C

eggs, meat, feathers, aesthetic qualities, or companionship. The word "chicken" comes from the Old English word *cycen*, meaning young fowl, which in Middle English evolved to mean a young chicken and eventually came to mean a chicken of any age.

The taxonomic classification of a chicken is this: kingdom *Animalia* (all animals); phylum *Chordata* (having a spinal cord); subphylum *Vertebrata* (having a backbone); class *Aves* (all birds); order *Galliformes* (chickenlike birds); family *Phasianidae* (pheasant family); genus *Gallus* (chickens); species *domesticus* (domestic chickens). \ The meat of a chicken.

chicken anemia \ *See: infectious anemia*

chicken coop \ A shelter that houses chickens, which may be in the form of a shed, an ark, a hutch, a chicken tractor, or any number of other variations. An ideal chicken coop has these features:

- Provides adequate space for the number of chickens
- Is well ventilated
- Is free of drafts
- Maintains a comfortable temperature year-round
- Protects the chickens from wind and sun
- Keeps out rodents, wild birds, and predatory animals
- Offers plenty of light during the day
- Has adequate roosting space for the number of birds
- Includes clean nests for the hens to lay eggs
- Has a sufficient number of sanitary feed and water stations
- Is easy to clean
- Provides access to the outdoors during the day
- Is located where drainage is good

[Also called: henhouse]

chicken diaper

chicken diaper \ A reusable plastic-lined diaper worn by a chicken kept in a house otherwise occupied by humans. Because chickens are difficult to housebreak, diapers are used for chickens kept as house pets, convalescing from an injury or illness, or being primped for show. Chicken diapers are available in various sizes, styles, and colors. An Internet search will reveal numerous sources, as well as detailed instructions on how to make your own.

chicken feed \ A ration fed to chickens that includes everything a chicken needs nutritionally to remain healthy. Commercial feeds are formulated according to age and life cycle; types include starter, grower, developer, finisher, breeder, and lay rations — not all of which are available in all areas. Exclusive of housing, feed accounts for about 70 percent of the cost of keeping chickens, making the subject of how to homegrow chicken feed a hot topic.

Chickens evolved in a forest environment that offered a broad variety of tasty things to eat. Confined chickens have the same basic nutritional needs, which include carbohydrates for energy (corn and other grains), protein (a combination of ingredients that together furnish all the amino acids needed to create a complete protein), and an array of vitamins and minerals that may be obtained from various supplements, as well as from sprouts, table scraps, and fresh greens.

chicken greens \ A mixture of greens grown specifically for feeding chickens to keep them healthy and ensure richer tasting deep-yellow yolks. Either chicken greens are cut and fed to the chickens or the birds are turned into the plot for a limited time each day to peck and scratch. Any mixture of lettuces, spinach, and other tender greens will delight chickens.

chicken mite \ *See: red mite*

chicken mobile home \ A portable shelter on wheels with an axle and designed to be moved by ATV, truck, trailer, or draft animal.
[Also called: eggmobile; henmobile]

chicken pox \ Avian pox, also called dry pox, a viral infection causing scabs on a chicken's skin that are similar, but unrelated, to the chicken pox affecting humans.

chicken rooster \ *See: bilateral gynandromorph*

chicken tractor \ A small, floorless, portable shelter for keeping chickens in a garden, typically on raised beds, so they will destroy weeds and weed seeds, eat cutworms and other pests, and fertilize the soil. To keep the chickens from foraging in their own droppings, the chicken tractor must be either moved often or bedded with deep (composting) litter. Although active chickens will scratch in the dirt, hence the name chicken tractor, standard hybrid broilers and other inactive types tend instead to compact the soil. In northern areas, the birds need alternative housing during rough winter weather. *[Also called: ark]*

chicken wire \ *See: hexagonal netting/wire*

chicken mobile home

cholesterol \ A kind of fat found in the bodies of both chickens and humans. Cholesterol is required for both the synthesis of vitamin D from sunshine and the production of sex hormones but can also collect in the bloodstream and clog the arteries. Many foods that contain cholesterol are also high in saturated fat.

An egg is the rare exception — it is high in cholesterol but contains little saturated fat. The yolk of one large egg contains about 185 milligrams of cholesterol and 1.5 grams of saturated fat — compared to a limit of 22 grams or less of saturated fat recommended for a person eating two thousand calories per day. Furthermore, eggs contain lecithin, which interferes with the body's absorption of cholesterol.

The cholesterol and saturated-fat content of eggs is influenced by the hen's diet. Pasturing layers, feeding them greens, or altering their ration to include 10 percent flaxseed can reduce both the cholesterol and the saturated-fat content of their eggs by some 25 percent. Because the eggs of hens that eat green plants have darker yolks than those of hens without access to greens, yolk color is a good reflection of an egg's lower levels of cholesterol and saturated fat.

The claim that eggs with blue or green shells are lower in cholesterol than white-shell eggs is false. In fact, eggs laid by hens of heavier breeds (including the Araucana) are likely to contain slightly more cholesterol than eggs laid by commercial Leghorn strains raised under similar circumstances, since the Leghorn produces more eggs and therefore has less time to put cholesterol into each one. However, the difference among breeds is so slight as to be insignificant.

chook \ Australian word for chicken.

chromosome \ A microscopic threadlike structure containing genes that carry hereditary determination. All the genetic information transmitted from a pair of chickens to their offspring is organized on chromosomes. Each chicken has 39 pair — 38 pair of autosomal chromosomes and one pair of sex chromosomes.

chronic respiratory disease (CRD) \ A bacterial infection of the respiratory system. Affected chickens typically survive but become carriers. This disease is therefore reportable in some states.

Cinnamon Queen \ Trade name for red sex-link hybrid chickens resulting from a cross between a New Hampshire cock and a silver laced Wyandotte hen.

circulated-air incubator \ An incubator with a built-in fan that constantly circulates air to maintain an adequate oxygen level and keep

circulated-air incubator

the temperature even throughout the incubator.
[Also called: fan-ventilated incubator; forced-air incubator]   *See also: incubator*

class \ A grouping of breeds for exhibition purposes. The American Poultry Association groups large breeds according to their place of origin: American, Asiatic, Continental (European, not including England), English, and Mediterranean. A catchall class is All Other Standard Breeds (AOSB), which includes Oriental breeds and any that don't readily fit into another class. Each large breed is listed in only one class.

Bantams are grouped into five classes, one of which is Game. The other four relate to comb style and leg feathering: Single Comb Clean Leg (SCCL), Rose Comb Clean Leg (RCCL), All Other Combed Clean Leg (AOCCL), and Feather Leg (FL). The same breed may be represented in different classes, depending on whether the comb is single or rose.

The American Bantam Association's classes are similar but include some breeds not recognized by the American Poultry Association and omit some of those the APA recognizes. Additionally, the ABA has two game classes: one for Modern Game and another combining American Game and Old English Game. \ A grouping of chickens competing against each other at a show, all of the same breed, variety, and age. Most shows have four age groups for each variety: cock, hen, cockerel, pullet. The terms "cock" and "hen" generally refer to birds one year of age or older, but some shows define them as having hatched before

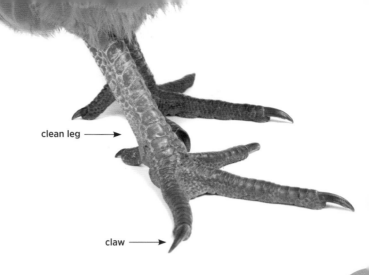

clean leg ⟶

claw ⟶

January first of the year of the show. The terms "cockerel" and "pullet" refer to birds that are less than one year of age, although some shows define them as having hatched during the calendar year the show is held. If you will be showing young chickens, check the definitions in the premium list to make sure your birds are entered in the proper classes.

clavel/clavell comb \ A carnation comb. The Spanish word *clavel* derives from the Old Catalan word *clavell*, meaning carnation.   *See also: carnation comb; comb styles*

claw \ The curved, pointed horny nail at the end of each toe.
[Also called: toenail]

clean face \ Characteristic of an Araucana that lacks ear tufts.
[Also called: clean head]

clean head \ *See: clean face*

clean leg \ Having no feathers growing on the legs or feet.

clerk \ A show judge's assistant, whose job is to locate each class to be judged; let the judge know how many birds are in each class so none will be missed if the class continues around a corner or across the aisle; keep track of the judge's *Standard* and other paraphernalia that might otherwise be misplaced; prevent bystanders from interrupting while the class is being judged; record the judge's placings; and return forms to the show secretary so ribbons and other awards may be distributed without delay.

cloaca \ The chamber just inside the vent where the digestive, excretory, and reproductive tracts come together. When an egg is ready to be laid, the shell gland at the bottom end of the oviduct remains wrapped around the egg while it passes through the cloaca, pressing shut the intestinal opening, so a hen cannot lay an egg and poop at the same time.

cloacal bursa \ A blind sac attached to the top of the cloaca enclosing tissue arranged into a series of leaflike folds that are filled with blood vessels and lymph nodules. In a chick this organ activates the production of antibodies and controls immunity by releasing so-called b-cells (bursa-derived cells) into the bird's body as it matures. By the time the chicken reaches sexual maturity, the cloacal bursa has atrophied. Some viral diseases, most commonly Marek's disease in backyard flocks and infectious bursal disease in commercial chickens, cause premature atrophy that permanently compromises the bird's immune system. *[Also called: bursa of Fabricius]*

cloacal sexing \ *See: vent sexing*

cloacitis \ An uncommon infection consisting of a chronic inflammation of the cloaca, characterized by a smelly, loose discharge sticking to the feathers around a swollen, reddened vent. Cloacitis is not a specific disease but a condition that has many different causes. It is typically triggered by severe stress that increases intestinal pH and compromises the immune system, opening the way to an infection of one kind or another. *[Also called: vent gleet (gleet is an old term for any abnormal discharge)]*

closed flock \ A population of chickens that have no contact with other chickens and into which no chickens are introduced from outside sources. Any additions to the flock are hatched from eggs fertilized and laid within the flock. The primary purpose of maintaining a closed flock is to prevent the introduction of diseases.

closed show \ A show at which either the coop tags are folded or exhibitors are identified by numbers, so the judge can't see the name of a bird's owner and be influenced by the owner's reputation.

close feathered \ Description of a breed with feathers that lie close to the body, as opposed to loose-feathered breeds with feathers that tend to angle away from the body. For example, Rhode

digestive
anatomy

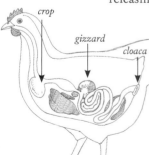

crop

gizzard

cloaca

Island Reds, New Hampshires, and Plymouth Rocks hold their feathers more tightly than the loosely held feathers of Cochins, Orpingtons, and Silkies. *[Also called: tight feathered]*

clown-faced chicken \ Spanish white-faced black chicken. Its prominent white face and large red wattles are reminiscent of a painted clown. *See also: Spanish*

clubbed down \ Clumped or matted down on an embryo or newly hatched chick that remains encased in down sheaths because the sheaths fail to rupture properly, causing the down to curl instead of fluff out. The condition is often associated with depressed hatchability but also occurs in chicks from a normal hatch. It is sometimes attributable to riboflavin (vitamin B_2) deficiency in the breeder flock, but in other cases the cause remains undetermined.

cluck \ A short, low-pitched, repetitive sound made by a hen to comfort her chicks and keep them from straying. Sometimes a setting hen starts clucking well before her eggs are due to hatch, especially when she leaves the nest briefly to eat or eliminate. Most hens start clucking when their chicks start peeping just prior to hatching. \ A hen with chicks.

clucker \ A hen with chicks.

clutch \ A batch of eggs that are hatched together, from the Old Norse word *klekja*, meaning to hatch. *See also: setting* \ All the eggs a hen produces in one laying cycle, which in modern domesticated chickens exceeds the number of eggs the hen can reasonably hatch at one time. \ A brood of chicks.

cobby \ Short, round, and compact, like the characteristic shape of Wyandottes, Sebrights, and Japanese bantams.

cocci/coxy \ *See: coccidiosis.*

coccidia \ Several species of parasitic protozoa, most belonging to the genus *Eimeria*, that infect a chicken. Each species invades a different part of the intestine, and a bird may be infected by more than one species at a time. The species of coccidia that infect chickens do not affect other kinds of livestock, and chickens cannot be infected by any of the many species of coccidia that affect other animals, including other poultry species.

clown-faced chicken

coccidiasis \ The condition of being infected with coccidia without showing any signs. Infected chickens that appear healthy shed billions of coccidia eggs that readily infect younger birds.

coccidiocide \ One of several drugs used to treat coccidiosis by destroying coccidia during their development. A coccidiocide is generally dissolved in drinking water, since infected chickens may stop eating but will continue to drink. Not all coccidiocides work against all types of coccidia, and using the wrong drug can do more harm than good. At the first sign of coccidiosis, take a sample of fresh droppings to a veterinarian and ask for a fecal test to find out what kind of coccidia are involved and therefore which medication to use. Proper

COCCIDIOSIS

Being infected with coccidia results in an intestinal disease that interferes with nutrient absorption. The main signs of coccidiosis are slow growth and loose, watery, or off-color droppings. If blood appears in the droppings, the illness is serious — birds may survive but are unlikely to thrive. The disease may develop gradually, or bloody diarrhea and death may come on quickly. Among young chickens it results in failure to grow at a normal rate and is the most common cause of death. In mature birds the chief sign is a decrease in laying.

Coccidiosis usually occurs in damp, unclean, overcrowded housing and is most common in growing birds three to six weeks of age, with the worst cases appearing at four to five weeks. In a properly clean environment, floor-raised chickens develop immunity through gradual exposure as they mature. Chicks raised on wire and later moved to floor housing have had little exposure and therefore have no immunity and can become seriously infected. Even mature birds can become infected in hot, humid weather when coccidia proliferate rapidly.

Chickens become immune in two ways: through gradual exposure or by surviving the illness. But they become immune only to the species occurring in their environment. Healthy chickens brought together from different sources may not all be immune to the same species and therefore may transmit coccidiosis to one another — with devastating consequences.

This serious disease may be prevented by the use of an anticoccidial vaccine or by using a probiotic to encourage competitive exclusion. It may be controlled with a coccidiostat or treated with a coccidiocide. If you prefer to avoid the use of drugs for chicks raised in a warm, humid climate or at a warm, humid time of year, you must maintain meticulous sanitation of the brooding facility and get the birds into a pasture rotation as quickly as possible — no later than three to four weeks of age. In all cases, drinking water must be kept free of droppings.

[Also called: cocci; coxy; Eimeria infection]

treatment, started early, works quickly and effectively.

coccidiostat \ One of several drugs used to inhibit the development of coccidiosis in chicks. The coccidiostat may be added to drinking water or, more commonly, incorporated into starter ration (medicated feed), since chicks that get overheated (in warm weather or in a too-hot brooder) may drink more than usual and obtain a toxic dose. A coccidiostat will not cure chicks already infected with coccidiosis.

Cochin \ A chicken originating in Asia as a large meat breed and developed in England as an ornamental breed. The name Cochin comes from its place of origin, the Cochinchina region of southern Vietnam. The loose feathering, combined with a puffy tail, feathered legs, and an abundance of down, gives these chickens the appearance of being spherical and also makes them winter hardy. Cochins have a single comb, come in numerous color varieties, and may be either large or bantam. The hens lay eggs with brown shells and easily go broody.

cock \ A mature male chicken, generally considered to be at least one year old. A hen will lay eggs whether or not a cock is around, but a cock is needed for the eggs to be fertile and able to hatch. *[Also called: rooster (even though both genders roost)]*

cock-a-doodle-doo \ An imitation of the sound a cock makes, from the Irish Gaelic *cuc-a-dudal-du.* Cocks don't always follow the plan. Some issue an inverted a-cock-doodle-doo, others a shortened cock-doodle-doo.

cock/cockerel breeder \ A cock or hen selected to produce outstanding standard-bred cockerels for exhibition. For some varieties, achieving exhibition quality requires breeding different strains to obtain females and males that conform to the standard for their variety. The partridge Wyandotte is one example. To obtain an exhibition-quality cock with the requisite solid black breast requires breeding females that lack the fine penciling of an exhibition-quality hen. *See also: pullet breeder*

cock egg \ A yolkless egg. It is sometimes called a cock egg because people once believed an egg without a yolk was laid by a rooster. *See also: yolkless egg*

Cochin hen

cock fighting
(TOP):
*natural
sparring among
roosters to
determine
status*

cockfighting
(BOTTOM):
*deliberate
maximization
of aggressive
tendencies for
the amusement
of handlers*

cocker \ A person who maintains fighting cocks.

cockerel \ A feathered-out male chicken under one year of age. At the age of between three and eight weeks, depending on breed, cockerels start developing reddened combs and wattles. Unless they're Sebrights or Campines, which are hen-feathered breeds, the males soon thereafter develop pointed back and saddle feathers and long tail sickles, in contrast to the more rounded back and saddle feathers and shorter tails of pullets. At about the same time, peck-order fighting gets serious and sexual activity starts. At that stage, cockerels should be separated from pullets or at least reduced in number to a reasonable ratio for the number of pullets.

cock fighting \ Sparring among roosters to determine peck-order status. Some individual cocks, and some entire breeds, have a greater inclination to fight. Roosters in a typical backyard flock will fight only until one backs down, but in some cases two cocks may fight until one kills the other.

Cocks are less apt to constantly challenge each other if you keep more than two and if their housing offers sufficient room for those lower in the pecking order to get away from the dominant ones. When using several cocks for flock breeding, keep them in two or more groups and rotate the groups, rather than rotating individuals — the less you disturb the peck order, the less fighting will occur. During breeding season, if you lose one cock out of a group, a possible slight drop in fertility is better than having fertility plummet because of peck-order fighting caused by introducing a new cock.

To further minimize fighting:
- Furnish one feeding station for each rooster around which he can gather his harem of hens
- Space the feeders well apart
- Locate them where each rooster can reach his station without passing through another's feeder territory.

cockfighting \ The deliberate maximization of the aggressive tendencies of roosters for the amusement of their handlers. Cockfighting may have played an important role in the early domestication of chickens, and certain bloodlines are still bred for the cocks' aggressiveness toward other cocks.

This behavior is more typical of the game breeds that have been around for centuries, although even among these breeds the strains bred for exhibition have had much of the fight bred out of them. The characteristics of a fighting breed include a big-boned body with heavy muscling (for strength), long neck and legs (for reach), hard feathering (for armor), and a hardy constitution (for resilience). The cocks are fed a specialized diet and given regular exercise. The result is a lean, tough, sinewy bird.

Cocks in the fighting pit are paired by weight and have been trained not to back down but to keep fighting, sometimes to the death. To make matters worse, the cocks are fitted with razorlike spurs to augment their already formidable natural spurs. Cockfighting, once one of America's national sports, is now considered inhumane and is illegal in most of the United States and its territories. *[Also called: cocking]*

cocking \ *See: cockfighting*

coelom \ *See: abdomen*

colibacillosis \ A complex group of infectious diseases caused by one or more strains of *Escherichia coli*, the so-called coliform bacteria, that normally live in the intestines of chickens. Most strains do not cause disease, but some are opportunistic and can cause infections that range from severe and acute to mild and chronic.

The susceptibility of chickens varies with the birds' strain and age — infection is more common in young birds than in older ones. The bacteria are transmitted through the shells of hatching eggs, causing dead embryos or chicks with infected navels (omphalitis). In the brooder *E. coli* is spread rapidly by means of infected droppings picked from litter, feed, or water.

The bacteria enter a chicken by way of either the digestive or the respiratory system but may eventually settle in the eye, heart, liver, navel, oviduct, leg joints, or wing joints. They may get into the bloodstream, causing acute septicemia and the sudden death of an apparently healthy chicken. Respiratory colibacillosis may be prevented by minimizing stress, providing good ventilation, and keeping chickens free of other infections. Intestinal colibacillosis is more difficult to avoid but may be minimized by controlling rodents and keeping feed and drinking water free of chicken droppings.

cockerel

coliform bacteria \ *See: Escherichia coli.*

color \ Not just the appearance of the chicken's visible plumage but also the color of the underfluff and the skin underlying the plumage, as well as the shanks, toes, comb, earlobe, eye, and skin surrounding the eye. Each standardized breed is unique in the color combination of its various parts.

comb anatomy
single (LEFT) and rose (RIGHT)

undesirable combs

colored egg \ A blue or green egg, such as is produced by an Ameraucana, Araucana, or Easter Egger.

color sexing \ Taking advantage of the sex-link gene that controls feather color. For instance, Barred Plymouth Rock chicks all look pretty much alike — black with white spots on the head — but the spots on the cockerels' heads tend to be irregular and scattered, while the pullets' spots are more compact and round. Color sexing is commonly used to produce hybrid brown-egg layers known as black sex-links (which takes advantage of the Barred Rocks' head spots by breeding them only on cockerels) and red sex-links (which uses other color factors to distinguish cockerels from pullets).

comb \ A fleshy crownlike protrusion on top of a chicken's head, usually more prominent in cocks than in hens. Most breeds have bright red combs. Sebrights, Sumatras, and some Modern Games have purplish combs, and a Silkie's comb is so dark it's nearly black.

comb, undesirable \ A comb that significantly deviates from the standard for the breed. Lopped comb is desirable in New Hampshire and Mediterranean hens but undesirable in most other breeds. Always undesirable are: hollow comb, inverted spike, side sprig, split comb, telescope comb, thumb marks, and twisted comb.

commercial ration \ Prepackaged chicken feed intended to provide all of a chicken's nutritional needs at various stages of life and life cycle. Rations for

COMB STYLES

Combs come in many styles, but not everyone agrees on how many styles exist or what each style is called. The single comb is by far the most common style. Other styles include buttercup, carnation, cushion, pea, rose, strawberry, V, and walnut combs. Most breeds are characterized by a specific comb style. Ancona, Dorking, Leghorn, Minorca, Nankin, Rhode Island Red, and Rhode Island White are exceptions; these breeds have both single-comb and rose-comb varieties.

buttercup

carnation

cushion

pea

rose, nonspiked

rose, spiked

single

strawberry

V

walnut

chicks contain a high amount of protein. As birds grow, they gradually need less protein and more energy. Commercial rations are therefore formulated according to age and life cycle, and include starter, grower, developer, and lay ration. Meat birds have their own formulations for starter/grower ration and finisher ration intended to induce rapid growth. Layer ration may be available with different levels of protein, the higher levels used both during hot weather when hens tend to eat less and also to improve the hatchability of eggs collected for incubation. Commercial rations come in three basic forms: mash, pellets, and crumbles. The variety of choices in form and formula varies and is greatest in areas where chickens are more numerous.

Most formulas include corn for energy and soybean meal for protein, despite the fact that soy meal may contain solvent residues left after the oil has been extracted and, further, has inherent properties that inhibit the absorption of other nutrients. Another issue of concern with any premixed ration is that one size fits all — fixed nutrients mean an individual chicken may get more of some nutrients than it needs while it's trying to satisfy its requirements for others. A chicken fed a wider variety of feedstuffs, and especially with access to forage, will pick and choose what it needs to obtain a balanced diet.

competitive exclusion \

Encouraging the growth of beneficial gut flora to fend off harmful organisms, the concept behind the use of probiotics and similar products. Beneficial bacteria may be encouraged by reducing intestinal pH.

On a commercial scale, pH is reduced by adjusting the feed formula to be more acidic. For the home flock, acidity may be introduced by adding apple cider vinegar to the drinking water, which reduces crop pH and thereby increases the number of beneficial bacteria that make it to the intestines.   *See also: vinegar*

complete feed \ A commercial
ration, which supposedly contains every nutrient a chicken needs to maintain proper growth and health.

complete protein \ Description
of a ration containing all the essential amino acids.

compost \ A combination of ani-
mal matter, such as chicken droppings, and plant matter, such as used bedding, that has decomposed into a rich black humus applied to soil to improve its fertility. Fresh chicken manure is so high in nitrogen it can literally burn growing plants. Composting not only transforms the manure's nitrogen into a form that's readily usable by plants without causing damage, but it also destroys bacteria, viruses, coccidia, and parasitic worm eggs. A carbon-containing substance — such as shavings, straw, weeds, or grass clippings — is needed to prevent the aging manure from overheating and drying into a powdery white, nutrient-poor, ashy substance called fire fang.

The combination of aged manure and bedding has average nitrogen (N), phosphate (P), and potash (K) values of 1.8, 1.4, and 0.8, respectively (as percentages of total weight). A good minimum yearly application is

C

45 pounds — approximately the amount produced by one lightweight layer in a year — per 100 square feet of garden (in approximate metric equivalents, a minimum application on a 10 sq m garden is 23 kg). In practical terms, that's about a ¾-inch (2 cm) layer of chicken-litter compost spread over 100 square feet (9.3 sq m) of garden soil.

composting litter \ Chicken-coop bedding that has sufficient depth to allow natural composting to occur, in the process stabilizing or fixing both nitrogen and potash. Start in spring with 4 inches (10 cm) of clean bedding. Whenever the surface gets packed or matted, break it up and stir in a little fresh bedding, enough to absorb prevailing moisture. The bedding should be 10 inches (25 cm) deep by the start of winter. Continue adding fresh litter as needed to absorb the amount of manure deposited.

Composting litter reduces in volume like a compost pile. Once the litter reaches enough volume to start actively composting, the amount of volume reduction should roughly equal the amount of new litter added — 12 to 15 inches (30 to 38 cm) of depth should strike the right balance. Rake or stir the bedding as often as necessary to keep the surface from crusting over, and remove manure accumulating beneath perches (or collect it in a droppings pit).

Adequate ventilation is needed to ensure the litter retains the right amount of moisture, for good fermentation. To test litter moisture, pick up a handful and squeeze. If the moisture level is just right, the litter will stick slightly to your hand but will break up when you let go. If

it's too dry, it won't stick to your hand; if it's too wet, it will ball up and not fall apart easily when you drop it. Too-dry litter is exceedingly dusty; too-damp litter releases ammonia fumes that irritate avian and human eyes and respiratory tracts. Excessive moisture is more often a problem than excessive dryness, causing all manner of health issues.

After about six months, properly composting litter develops sanitizing properties, and the heat produced by fermentation will keep chickens warm during the cool months. In warmer months flies are less of a problem because accumulated dry manure attracts natural fly predators and parasites.

A flock may be kept on the same composting litter for years, provided the bedding doesn't get damp, remains warm, and no serious disease breaks out. If the nitrogen-rich bedding is needed to fertilize a garden, or the composting litter makes the coop too warm in a hot climate, remove the litter each spring and start summer with fresh, cool bedding. *[Also called: built-up litter]*

concave sweep \ A characteristic of those breeds with a back that curves continuously from the shoulders to partway up the tail. The sweep may be slight (as in Plymouth Rocks), gradual (as in New Hampshires), short (as in Jersey Giants), sharp (as in Langshans), or rising (as in Anconas).

condition \ A chicken's state of health and cleanliness. A chicken in top condition has that undefinable glowing quality known as bloom.

concave
sweep

slight

short

sharp

conditioning \ The process of bringing a breeder or show bird to the peak of cleanliness and good health through good nutrition and grooming.

confinement \ A housing method in which chickens are kept within fixed quarters (indoor confinement) or movable quarters (range confinement) offering protection from inclement weather and predation. Confinement indoors without adequate ventilation carries the risk of respiratory illness due to ammonia from droppings and dust from litter, skin, and feathers. Confinement on range requires frequent moving of the shelter to prevent an unhealthful buildup of mud and manure.

conformation \ A chicken's body shape. The conformation of many breeds found in North America is standardized in the books *American Standard of Perfection* and *Bantam Standard*. Standards adopted by other countries may differ.

Continental class \ One of six groupings into which the American Poultry Association organizes large chicken breeds. The breeds in this class originated primarily in Europe (excluding England). This class is further subdivided into Northern European, Polish, and French. In the Northern European group the Campine is from Belgium; the Lakenvelder from Germany; the Barnevelder, Hamburg, and Welsumer from Holland. The Polish, too, is from Holland (its name deriving from the knob or pole on top of its skull) and is further subdivided into two breeds according to whether the birds are bearded or nonbearded. The Continental breeds from France are Crevecoeur, Faverolle, Houdan, and La Fleche.

continuous hatching \ Continuously putting fresh eggs into an incubator as the already incubated eggs hatch. Since the incubation of chicken eggs takes 21 days, a common practice is to add eggs once a week so that, after the first three weeks, a new batch of chicks hatches weekly. To avoid contamination resulting from eggs hatching in the incubator itself, moving the about-to-hatch eggs to a hatcher is an important sanitation measure.

contour feathers \ The outermost, webbed feathers covering a chicken's body and protecting it from sun, wind, rain, and injury.

controlled lighting \ Manipulating the number of hours, generally by means of a timer, that chicks or chickens are housed under light (electric and natural combined) to achieve desired results. A reflector behind each bulb increases its intensity, allowing you to use less wattage than you would otherwise need. Dust and cobwebs accumulating on bulbs decrease light intensity. To maintain the effectiveness of your controlled-lighting program, dust the bulbs weekly and replace any that burn out.

When using compact fluorescent bulbs, use soft white (warm-wavelength) lights, since cool-wavelength lights are less effective for chickens. If your coop is not outfitted with electricity, you can provide lighting with 12-volt bulbs designed for recreational vehicles, powered by a battery connected to a solar recharger. Since a timer would drain too much power from your battery, 12-volt lights either must be left on all the time or manually turned on and off.

[*Also called: photostimulation*]

See also: controlled lighting for broilers; controlled lighting for cocks; controlled lighting for hens; controlled lighting for pullets

controlled lighting for broilers \ How many total light hours

work best for meat birds is a matter of debate. The trend in commercial production is to shorten daylight hours for chicks 2 to 14 days old, giving them less time to eat, which slows their growth rate and thereby reduces leg problems and other complications resulting from too-rapid growth. After the broilers reach 2 weeks of age, light hours are increased to encourage them to eat and grow, with just enough light to let the birds find feeders but not enough to inspire them to engage in other activities that would burn off energy. Under this program, chicks are started under 60-watt incandescent bulbs, or 15-watt compact fluorescents, placed in reflectors 7 to 8 feet (2 to 2.5 m) above the floor. After two weeks, switch to 15-watt incandescent bulbs, or 5-watt compact fluorescents. Allow one bulb-watt per 8 square feet (0.75 sq m) of living space.

Broilers raised under continuous light may panic, pile up, and smother if the power fails. To get them used to lights-out, turn lights off for at least 1 hour during the night. For efficient growth, lights need not be on more than 14 hours per day. In hot weather, which makes chickens lethargic, dividing the light hours between morning and evening encourages chicks to eat while the temperature is cooler.

MAXIMUM HEIGHT OF LIGHTS ABOVE BIRDS

Incandescent/ fluorescent	AGE 12–21 WEEKS		AGE 21+ WEEKS	
	With reflector	No reflector	With reflector	No reflector
15/3 WATTS	5.0' (1.5m)	3.5' (1.0m)	3.5' (1.0m)	2.5' (.75m)
25/5 WATTS	6.5' (2.0m)	4.5' (1.5m)	4.5' (1.5m)	3.0' (1.0m)
40/10 WATTS	9.0' (3.0m)	6.5' (2.0m)	6.5' (2.0m)	4.5' (1.5m)
60/15 WATTS	14.0' (4.0m)	10.0' (3.0m)	10.0' (3.0m)	7.0' (2.0m)
75/20 WATTS	15.5' (5.0m)	10.5' (3.25m)	10.5' (3.25m)	7.5' (2.25m)

Arrange fixtures to minimize shadows, and space them no farther apart than 1.5 times their height above the birds.

Adapted from: *Storey's Guide to Raising Chickens,* 3rd edition

controlled lighting for cocks

\ Early in the year, when days are short, semen production may be stimulated and fertility improved by increasing a cock's exposure to light. Furnish at least 14 hours of light, artificial light and daylight combined, starting 4 to 6 weeks prior to the collection of eggs for hatching. The controlled lighting setup for cocks is the same as for laying hens.

controlled lighting for hens

\ In the natural course of events, pullets hatch in spring as daylight hours are increasing, and mature during summer and autumn as daylight hours decrease, often laying through the winter of their first year. The following spring, as day length again begins to increase, they start a new reproductive cycle. As hens they continue to lay until either the number of light hours per day or the degree of light intensity signals the end of the reproductive cycle.

Most hens therefore stop laying during the shorter days of winter. Using lights to compensate for decreasing amounts of natural daylight tricks them into thinking the season remains right for reproduction. Start augmenting natural light when day lengths approach 15 hours, which in most parts of the United States occurs in September. Continue the lighting program throughout the winter and into spring, until natural daylight is back up to 15 hours per day.

One 60-watt incandescent bulb, or a 15-watt compact fluorescent, 7 feet (0.2 m) above the floor, provides enough light for about 200 square feet (18.5 sq m) of living space. Place the bulb in the center of the area to be lighted, preferably over feeders and away from the nesting area. If your shelter is so large you need more than one fixture, the distance between them should be no more than 1.5 times their height above the birds. Arrange multiple fixtures to minimize shadows, except over the nesting area, which should remain darkened.

Leaving lights on all the time is not only wasteful but encourages hens to spend too much time indoors during the day. It also does not give them the daily 6 to 8 hours of restful darkness they need to maintain good immunity. On the other hand, if the lights go off for one entire day, your hens may go into a molt and stop laying. Install multiple lights in case one burns out, using a timer to ensure the hens won't be left in the dark. Set the timer to turn lights on for a few hours at the same time every morning and again for a few hours in the evening, to bracket the changing daylight hours and simulate a constant 15-hour day.

controlled lighting for pullets

\ During their first year, pullets may lay throughout the winter without the benefit of controlled lighting. However, controlled lighting may be needed to delay maturity. Under normal circumstances pullets reach maturity during the season of decreasing day length. If you raise pullets between August and March, the increasing day length that normally triggers reproduction will speed up their maturity, the more so as they approach laying age. Pullets that start laying before their bodies are ready will lay smaller eggs and fewer of them and are more likely to suffer reproductive issues such as prolapse.

Pullets should be raised in either constant 8- to 10-hour days or in decreasing light. Consult an almanac to determine how long the sun will be up on days occurring 24 weeks from the date the pullets hatch. Add 6 hours to that day length, and start your pullet chicks under that amount of light (daylight and electric combined). Reduce the total lighting by 15 minutes each week, bringing your pullets to a 14-hour day by the time they start laying. When they reach 24 weeks of age, add 30 minutes per week for 2 weeks to increase total day length to 15 hours.

Since spring is the natural season for chicks to hatch, pullets hatched from April through July and raised in natural light will mature at the normal rate, making them less likely to experience reproductive issues.

controlled molt \ *See: forced molt*

coop \ *See: chicken coop; show coop.*

Cooperative Extension System \ A nationwide, non-credit educational network funded by the National Institute of Food and Agriculture. Each state and territory of the United States has an Extension office at its land-grant university, along with a network of local or regional offices, typically at the county level. The offices are staffed by Extension agents, who provide useful, practical, and research-based information related to a variety of rural subjects, including raising chickens. Extension also sponsors the youth development program 4-H and its many activities for young people interested in keeping chickens.

coop tag \ A card attached to each cage at a chicken show bearing information that identifies the bird assigned to that coop and the bird's owner. The card is used by the show judge to note the bird's placement in relation to others in its class. A thoughtful judge may comment on the bird's outstanding good or bad features. Not all coop tags are alike, as they are available variously from the American Bantam Association, the American Poultry Association, and feed manufacturers; some poultry clubs even design and print their own.

| VARIETY | |
| BREED | |

ENTRY NO.	BAND NO.
Cock	Pullet
Hen	Young Trio
Cockerel	Old Trio

| Judge's Section |

coop tag

cooping in \ Bringing exhibition chickens into a showroom at the appointed time, placing them in their assigned show coops, and making sure they have adequate bedding, water, and feed. Show birds are organized according to classification and age; all the birds to be judged against one another are displayed in adjacent coops along one aisle (or in adjoining aisles, if the class is large).

cooping out \ Removing exhibition chickens from the showroom after all prizes have been awarded and the show is over. Some shows have strong rules against the practice of cooping out early, such as losing eligibility to enter the next show. Having a bird disqualified is not sufficient reason for early coop out. Neither is selling a bird to someone who's eager to take it home. A legitimate reason for cooping out early is to remove an ill or injured bird from the showroom.

coop size \ *See: space requirements*

coop training \ Teaching an exhibition chicken what to expect and what's expected of it at a show. Training helps a bird show to its best advantage by getting it used to being close to people and being handled, and ensures the bird won't be disoriented by unfamiliar surroundings. Training also minimizes stress, thus reducing the chance the bird will experience reduced immunity as a result of being shown.

Cornish \ A breed developed in and around Cornwall, England, by crossing Aseels with other game breeds, originally for cockfighting, later for meat production and exhibition. Though the Cornish name is derived from Cornwall, in England the Cornish is known as Indian Game. These chickens are hard feathered, have a pea comb, come in several varieties, and may be large or bantam. Strains of the white variety in the large size are used to develop the hybrid chickens favored for industrial meat production. The exaggerated width of the cock's breast sets its short legs so far apart that the straightbred Cornish rooster cannot breed naturally; eggs for incubation are obtained by artificial insemination. Cornish hens are unusual in having identical conformation to the roosters'. The hens lay eggs with tinted shells and make good broodies.

Cornish cross \ A broiler developed by crossing Cornish with some other breed, most commonly a Plymouth Rock.

Cornish game hen \ Not a game breed, and not a hen, but a Cornish Rock broiler pullet butchered at three or four weeks of age as a 1- to 2-pound (0.5 to 0.9 kg) single serving.

Cornish Rock broiler \ Broilers commercially developed by crossing specialized strains of white Cornish cocks with equally specialized strains of white Plymouth Rock hens. Cornish-cross broilers are interested in only one thing: eating. Commercial broilers grow fast and tender, reaching 5 pounds (2.25 kg) in 6 to 7 weeks. When they're not eating, they sit around developing bad habits, such as picking on each other or growing so big so fast they develop bone ailments or heart failure. Because they were developed to be raised in climate-controlled

Cornish

housing, they don't actively forage and won't do well outdoors when the weather is extremely hot or cold. Managing these hybrids therefore requires careful attention.

A commercial meat bird eats just 2 pounds (0.9 kg) of feed for each pound (0.45 kg) of weight gained. A hybrid layer, by comparison, eats three to five times as much for the same weight gain. Raising a hybrid meat flock is a short-term project. You buy a batch of Cornish-cross chicks, feed them to butchering age, dispatch them into the freezer, and enjoy the fruits of your labor for the rest of the year. Since these birds aren't around long, their ability to grow quickly on the least possible amount of feed takes precedence over fancy appearance.

coryza \ Infectious coryza, the chicken's equivalent of a cold.

courtship call/croon \ The low sound a cock makes while doing the courtship dance.

courtship dance \ The display a cock makes while trying to attract a hen by circling her while flicking one wing against the ground and singing the courtship croon.

coverts \ Contour feathers that cover the main tail feathers (tail coverts) and the bases of the wings' flight feathers (wing coverts).
See page 115 for illustration.

coxy \ *See: coccidiosis*

craw \ *See: crop*

creamy \ The yellowish tinge characteristic of newly emerging white feathers, as distinct from brassy mature feathers.

C

Cornish Rock

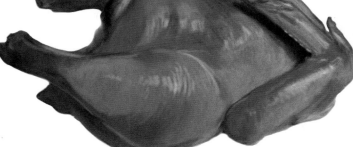
Cornish game hen

creeper \ Any chicken exhibiting the creeper gene, so called because it appears to creep, slide, or float as it moves around.

Creeper \ A bantam breed with shanks shorter than 1½ inches (3.8 cm), once recognized by the American Bantam Association in black, cuckoo, silver, and white color varieties. This breed was likely related to the rare Japanese Creeper, known as Jitokko in Japan, although it lacked the beard, muffs, and crest characteristic of today's Jitokko.

crest

creeper gene \ A genetic trait that causes short legs in certain breeds, including Japanese and Creeper bantams, and results in a high rate of embryo deaths during the incubation of eggs produced from the mating of a cock and hen that both bear the lethal creeper trait. Chicks that inherit a short-leg allele from both parents die before hatching. Of the chicks that do hatch, one-third will inherit two long-leg alleles and be

Japanese
Creeper rooster
or Jitokko

unsuitable for showing or breeding, and the remaining two-thirds will inherit one long-leg and one short-leg allele, giving them the short legs that are desirable in a Japanese or other creeper bantam.

crest \ A spherical puff of feathers growing on the head of a breed that has a knob on top of its skull, as is characteristic of Crevecoeurs, Houdans, Polish, Silkies, and Sultans. A full crest restricts vision and hinders the bird's ability to get away from predators and to catch mates. *See also: anatomy* *[Also called: topknot]*

Crevecoeur \ A breed developed centuries ago in Normandy, France, and named after a village in that province. Originally, it was a dual-purpose farmstead breed but today is considered primarily ornamental. The Crevecoeur comes in a single color, black; has a crest, V-comb, and beard; and may be large or bantam. The hens lay a chalky white-shell egg and seldom brood.

crisscross inheritance \ The inheritance of a sex-link characteristic such that, in a given mating, all the pullets inherit the characteristic from the sire and all the cockerels inherit the corresponding characteristic from the dam. *See also: sex-link*

Croad Langshan \ Langshan. To distinguish it from other Langshan strains — including Germany's German Langshan and England's Modern Langshan — that

deviate from the original type, this breed is sometimes called the Croad Langshan. It was first exported from China by Major A. C. Croad of England. *See also: Langshan*

crooked neck \ *See: wry neck*

crop \ An expandable pouch at the base of a chicken's neck that bulges with feed after the bird has eaten and where digestive juices begin softening the feed before it moves into the chicken's stomach. **See page 60 for illustration.** *[Also called: craw]*
\ To trim or remove a bird's wattles, as is required for the exhibition of mature Modern or Old English Game cocks.

crop binding \ *See: crop impaction*

crop impaction \ The lodging of feed in the crop, which continues to swell and stop up the works so no feed can pass through. Crop impaction may occur when feed is withheld before worming, causing chickens to eat too much too fast afterward. A crop may also get impacted in a bird pastured where little is available to eat except long, tough, fibrous vegetation. The crop will feel tight and hard and can swell sufficiently to cut off the windpipe, suffocating the bird unless surgery is performed to remove the impaction. Crop impaction is unlikely to occur in properly fed birds.

crossbreed \ To mate a cock and hen of two different breeds. \ The offspring of a cock and hen of two different breeds. *See also: hybrid*

Crevecoeur hen

crossed beak \ A beak in which the upper and lower halves grow in opposite directions, preventing the chicken from pecking properly. Although the beak may usually be trimmed to bring the two halves together, in most cases a crossed beak is a genetic defect, and such a chicken should not be used for breeding. You will know it's genetic if the crossed beak appears almost from the time the chick hatches and keeps growing crossed no matter how often you trim it. A crossed beak may also occur in a growing or mature chicken that is not able to swipe its beak along the ground; for example, if the chicken is kept in a cage.

crossed beak

crow head

crow \ The characteristic call of a male chicken as an assertion of dominance, putting other males on notice while avoiding unnecessary fights. Some roosters crow more than others, and roosters that are within earshot of other roosters crow more than solo roosters. Cocks of the larger breeds have a deep crow that doesn't travel far, while smaller breeds have a high-pitched crow that travels a good distance.

The age at which cockerels start crowing depends on how rapidly they mature, which is partially breed dependent, since some breeds mature more quickly than others. A cock inherits his style of crowing, so all cocks within a given family sound somewhat similar, although each individual develops a distinctive variation.

Crowing occurs most frequently in the morning but may be heard at any time of day or night. During the dark of night, crowing is usually triggered by a sound or a suddenly appearing light the cock perceives as a potential intruder. Nighttime crowing may be minimized by putting shutters on coop windows to keep out passing lights or by leaving a light on in the coop. Softly playing a radio helps mask sounds coming from outside the coop, and insulating the coop walls and surrounding the building with shrubbery can muffle any crowing that does go on. Since a cock stretches his neck to crow, putting him overnight in a ventilated box or cage small enough to prevent a good stretch and covering the box to keep out light will discourage overnight crowing. \ To utter the sound characteristic of a cock.

crow head \ An undesirable head formation in which the head and beak are narrow, shallow, and overly refined; often accompanied by other significant defects.

crowing contest \ A competition to see whose rooster can crow the most within a certain period of time, usually ranging from 10 or 15 minutes to 30 minutes.

crowing hen \ A not particularly common phenomenon, except among longcrowers, in which case a crowing hen is considered valuable as a breeder. Among other breeds, in a flock that includes no roosters sometimes a hen takes on the masculine role, including crowing. An aging hen may crow during nonlaying periods, when male hormones exert greater influence than female hormones.

crowing
Ameraucana
rooster

Hens might also crow as a result of disease. A hen has two ovaries, but only the left one produces eggs, while the right one remains undeveloped. If the left ovary becomes inactive due to atrophy or disease, the testicular tissue of the right ovary is stimulated into functional activity, resulting in the hen's getting a dose of the male hormone responsible for crowing.

crown comb \ *See: carnation comb; comb styles*

crumbles \ Crushed pellets, fed to chicks that aren't big enough to swallow whole pellets and to mature chickens so they take longer to eat and as a result are less likely to get bored. The main disadvantage of crumbles is the waste that occurs when any are spilled.

Cubalaya \ The only breed developed in Cuba, intended as a dual-purpose chicken that is also suitable for cockfighting. The Cubalaya has a pea comb, comes in a few color varieties, and may be large or bantam. The cocks have short, knobby spurs (referred to as rosary spurs) or none at all. The rooster's sex feathers majestically flow together from the neck to the end of the tail. Both cocks and hens have a distinctive broad and downward-pointing tail, referred to as a lobster tail or shrimp tail. The hens lay eggs with tinted shells and make good broodies.

cuckoo

cuckoo \ The coarse and irregular barring of California Grays, Dominiques, barred Hollands, and any variety with plumage coloration designated as cuckoo. Some cuckoos, including cuckoo Marans and Dominiques, may be autosexed with fair accuracy. Cuckoo Marans cockerels have paler down than the pullets; Dominique cockerels have a larger or more scattered head spot than the pullets.

C

Cubalaya hen

Cubalaya rooster

cushion comb

cull \ A chicken that is removed from a flock because it is overly aggressive, diseased, deformed, nonproductive, or inferior in some other way. \ To eliminate an aggressive, diseased, deformed, nonproductive, or otherwise inferior chicken from a flock. In most cases cull means kill, but in some cases — such as that of an imperfect show chicken — the bird might suit someone desiring a pet rather than a breeder. A breeding flock can degenerate rapidly if no effort is made to select in favor of health, vigor, hardiness, temperament, and good reproduction.

cup comb \ *See: buttercup comb; comb styles*

curly \ The undesirable result of mating two frizzled chickens together. Twenty-five percent of the offspring will be curlies, having inherited the frizzle gene from both parents. Their sparse bristle feathers are so brittle they easily break away, leaving the chicken virtually bare. The curly thus lacks plumage protection against temperature extremes or sunburn.
[Also called: extreme frizzle; frazzle]

cushion \ An abundance of feathers over the back and base of the tail, giving the chicken the appearance of being round — characteristic of Cochins and Silkies but undesirable in most other breeds.

cushion comb \ A small, compact, low-growing comb that extends no farther back than the middle of the skull and is smooth and solid, with no depressions or spikes. This comb style is characteristic of the Chantecler. *See also: comb styles*

cuticle \ *See: bloom*

cutting \ A method of judging exhibition chickens that involves deducting points from a published general scale of points based on specific defects.

daily-move pen \ A portable, floorless shelter used to pasture chickens and moved to a fresh patch of grass each day.

dam \ Mother of a chick.

dam family \ Sibling chickens that have the same dam and sire.

dark meat/light meat \ Dark meat generally occurs in leg and thigh muscles that are heavily used; light meat occurs in breast muscles that get less exercise. A bird uses breast muscles for flying and leg muscles for walking, but chickens do more walking than flying. Active muscles need oxygen, and oxygen is carried in blood cells. The more active the muscles, the more blood they require, and the more blood they require, the darker the meat.

Muscles get their energy from fat stored within the muscle cells. The more the muscles are exercised, the more energy they need, so the more fat they store. The leg muscles thus contain more fat, which is why a chicken's hindquarters

have more flavor, while the less fatty breasts are considered more healthful. The breast of a fast-growing broiler has less fat and flavor than that of other breeds.

Dark meat is denser than light meat and therefore takes longer to cook. The older the chicken, the greater the difference between the density (and cooking time) of the active leg muscles and the less active breast muscles. Similarly, the meat of slow-growing, active breeds takes longer to cook than the meat of fast-growing, lethargic commercial-strain broilers.

daw eyed \ *See: pearl eyed*

day range \ *See: free range; pastured/ pasture raised; range feeding*

death among chickens \ The normal mortality rate among chickens is 5 percent per year; many small flocks experience far fewer. Occasionally finding a dead chicken does not necessarily mean some terrible disease is sweeping through your flock. Like humans, chickens can die at any age — sometimes suddenly — of natural causes. Finding several chickens dead within a short time, however, is reason for concern. Death may result from, among other things, degeneration of the intestine due to an enteric disease, blocking off of the airways due to a respiratory disease, inability to eat or breathe due to paralysis caused by a nervous disorder, lack of adequate feed or water, poisoning, or predation.

permanent
debeaking

temporary
debeaking

debeak \ To shorten a chicken's beak to prevent cannibalism among intensively raised hens, where the practice is euphemistically called beak trimming or beak conditioning. As a result of this permanent amputation, the hens cannot eat properly. Birds in a properly managed backyard flock should not need permanent debeaking. However, temporary debeaking may be considered for chicks that persistently peck each other and cannot be stopped. Nail clippers are used to remove one-fifth of the upper portion of the beak, which should grow back in about six weeks.

debrain \ To use a sticking knife to loosen the feathers of a slaughtered chicken that will be handpicked. After the chicken has been killed by slicing through its jugular vein, the knife is inserted into the chicken's mouth with a sharp edge oriented toward the groove at the roof. The knife is pushed toward the back of the skull and given a one-quarter twist. The trick is to avoid sticking the front of the brain, which causes feathers to tighten instead of loosen. A knife that hits home causes the dead bird to reflexively shudder and utter a characteristic squawk.
[*Also called: sticking*]

decrow \ To eliminate a cock's ability to crow, through a surgical procedure that usually involves cutting muscles that control the syrinx. Assuming you could find a vet to do it, the operation is expensive, risky (because the muscles are not easily accessible and lie near major blood vessels), and not always successful. The surgical procedure of caponizing minimizes crowing, but a capon is useless for breeding. Most people, including veterinarians, find the surgical procedures of decrowing and caponizing to be distasteful if not downright inhumane.

defect \ Any characteristic that makes a chicken less than perfect, as defined by the breed's or variety's standard description. When chickens compete at exhibition, the severity of a defect determines the number of points cut. A list of general defects applying to all breeds sets forth a value for each cut in relation to the total value established in the general scale of points. Additional defects apply to individual breeds or varieties. Serious defects, which are nearly all inherited, disqualify a chicken from receiving an award. These guidelines for exhibition birds are invaluable not only in evaluating potential show entries but also in selecting future breeding stock for any purpose.

If you raise meat birds for your own family use, you needn't be too concerned about freedom from defects. You might, in fact, cull defective birds from your flock by putting them into the freezer. But meat birds raised for market must be free of such defects as crooked breastbones, crooked or hunched backs, deformed legs and wings, bruises, cut or torn skin, breast blisters, and calluses.

defluorinated rock phosphate \ A mined mineral used as a dietary supplement. Some rock phosphate deposits contain a toxic level of fluorine, which is removed through defluorination.

D

dehydration \ Loss of body water. A chicken's body contains more than 50 percent water, and an egg is 65 percent water. A chicken's body cannot maintain proper function if the bird is deprived of water. Water deprivation may occur because the quantity is insufficient during hot weather, the water freezes during cold weather, or the water is contaminated or otherwise unpalatable. A loss of body water amounting to more than 12 percent results in death.

Delaware \ A breed developed in the state of Delaware from an off-color sport arising from broilers created by crossing Barred Plymouth Rock cocks with New Hampshire hens. The sport was bred back to New Hampshire hens to create a dual-purpose chicken having the sport's color — white plumage set off by slight barring in the neck, wing, and tail feathers. The Delaware has a single comb, comes in a single color, and may be large or bantam. The hens lay brown-shell eggs and make good broodies.

Delaware Blue Hen \ *See: blue hen's chicken*

depopulate \ To get rid of an entire flock, typically by killing, usually to prevent the spread of a disease.

depth of body \ *See: abdominal depth*

devil bird \ La Fleche. It is called the devil bird because of its solid black plumage and prominent bright red V-shaped comb reminiscent of the devil's horns.   *See also: La Fleche*

Delaware hen

D

dewlap \ A surplus of loose skin extending from beneath the back of a chicken's beak partway down the throat, as is characteristic of Brahma hens but undesirable in other breeds.

dewlap

deworm \ To treat chickens with a preparation designed to control intestinal parasitic worms. A healthy chicken can tolerate a certain amount of internal parasites. Deworming without knowing whether or not the chickens have a problem is not only a waste of money but may interfere with the chickens' ability to develop resistance to parasites. Rather than deworming just because everyone else does, have a veterinarian periodically examine fecal samples for signs of worms, then develop a deworming schedule based on your flock's specific need.
[*Also called: worming*]

devil bird

dewormer \ Any preparation designed to control intestinal parasitic worms. Dewormers are administered in one of two ways: by mouth (added to feed, water, or given directly by mouth), to be absorbed through the digestive tract; or injected, to be absorbed from tissue under a chicken's skin. Dewormers approved for poultry change from time to time; check with your state poultry specialist, veterinarian, or poultry products supplier regarding the latest regulations. Products come in varying strengths, and dosages vary accordingly, so always follow label directions.   *See also: anthelmintic*

diarrhea \ A condition in which a chicken poops often and in liquid form, staining the feathers below the vent. Diarrhea can lead to dehydration and possibly death, unless the chicken is rehydrated with electrolytes and water. The cause of the diarrhea must also be determined and treated. Possible causes include coccidiosis, a heavy worm infestation, or a bacterial infection.   *See also: cloacitis*

diatomaceous earth (DE) \ Diatom fossils ground into an abrasive powder and sometimes added to nests and dust baths to reduce body parasites. However, chickens are highly susceptible to respiratory problems, and routinely inhaling diatomaceous earth can make matters worse. But if the chickens or their facilities are infested with parasites, the benefits of using it may outweigh the dangers.

Diatomaceous earth is also sometimes fed to chickens as a dewormer, which supposedly causes dehydration and death to internal parasites. But when combined with a chicken's saliva, diatomaceous earth softens and loses its cutting edge. The only way it could dehydrate internal parasites would be if it contained a hydrophilic substance that draws moisture from a parasite's body, and such a substance would equally affect the chicken's innards. On the other hand, diatomaceous earth contains a lot of trace minerals that are beneficial to chickens, even if it does not act as a dewormer. *[Also called: fossil flour]*

dirties \ Eggs with shells that have more than one-quarter of the surface covered with dirt or stains. Eating such eggs may pose a health risk if they are not properly cleaned.

disease \ Any disorder of a chicken's body or bodily function, especially one producing specific signs or affecting a specific body part and not a direct result of injury. A disease may be infectious (caused by an invasion by some other organism) or noninfectious, meaning its origin is not biological. Infectious organisms of one sort or another are always in the environment but don't usually cause disease unless a flock is stressed by such things as crowding, unsanitary conditions, or changes in feed. Disease-causing organisms are spread through the air, soil, and water, as well as through contact with diseased chickens. They may be carried from flock to flock on the feet, fur, or feathers of other animals, especially rodents and wild birds, and on equipment, human clothing (particularly shoes), and vehicle tires.

The disease most likely to strike a specific group of chickens is influenced by the flock's purpose. Exhibition birds are most likely to get a respiratory disease that spreads through the air. Broilers are most likely to experience diseases related to nutrition and rapid growth, such as ascites or leg weaknesses. A breeder flock is more likely to experience a disease requiring long-term development, such as tuberculosis.

disease prevention \ Introducing new chickens to an existing flock always carries the risk of introducing a disease. Any newly acquired bird, or one returning from a show, should be isolated for at least two weeks and watched for signs of disease. To protect valuable genetic stock, a sacrificial chicken from the existing flock might be put into isolation with the new bird to determine if it is carrying a communicable disease that would be transmitted to the sacrificial bird.

Of all the various possible biosecurity measures, staying away from other flocks — and keeping other poultry people away from yours — is the most important one and the hardest one to observe. Most people who enjoy chickens like to visit other people with the same interest. To avoid spreading disease by stepping in contaminated manure, cover your shoes with disposable plastic bags before going into someone else's yard, and keep bags handy for visitors who come to your yard.

disease resistance \ The ability of chickens to resist diseases in their environment. The best ways to keep chickens healthy are to minimize stress in their environment and to breed for resistance by hatching eggs collected only from 100 percent healthy breeders. Keeping chickens for at least two years before using them as breeders gives you plenty of time to weed out any that might be susceptible to disease.

disease signs \ Unique signs by which a disease may be identified. Each disease also shares some signs with other diseases. Reduced egg production is often the first general sign of any disease,

CAUSES OF DISEASE

Infectious (invasion by another organism)	Noninfectious (nonbiological in origin)
Bacteria	Chemical poisoning
Mold and fungi	Hereditary defect
Parasites	Nutritional deficiency
Viruses	Unknown cause

From: *Storey's Guide to Raising Chickens*

soon accompanied by depression, list-lessness, hunching, hanging of the head, drooping wings, dull or ruffled feathers, loss of appetite, and weight loss or poor growth. Other signs to watch for include swelling in one or both eyes, increased thirst, and droppings with an unusual color, texture, or smell. A pattern of deaths is a sure sign of disease. Many diseases have such similar signs that the only way to get a positive diagnosis is by taking a few sick or recently dead birds to a pathology laboratory for analysis.

disease treatment \ The moment you suspect a chicken is sick, isolate it from the rest of the flock; then make a choice: Either humanely kill the bird and dispose of the body according to local health regulations (burn or deeply bury are two options), or try to find out what's ailing it and begin active treat-ment. Some diseases may be effectively treated if caught in the early stages and positively identified to allow initiation of a precise treatment. Bear in mind that:

- By the time you notice a chicken is sick, chances are it's too far gone to be treated effectively
- Keeping an unhealthy bird carries the risk that the afflicting disease will spread to others in your flock
- Chickens that survive a disease rarely reach their highest potential as layers, breeders, or show birds
- A chicken that gets a disease, even if it fully recovers, will pass its lack of resistance to its offspring
- A chicken that fully recovers may become a carrier, continuing to spread the disease without showing signs

- Some diseases are so serious the only way to stop them is to dispose of the entire flock and start over

disinfectant \ A chemical sanitizer. Use only disinfectants approved for use with poultry. Brand names include Germex, Tek-Trol, Oxine, and Vanodine. Alternatives are plain vinegar, or house-hold chlorine bleach combined with hot water at the rate of ¼ cup bleach per gallon (15 mL/L) of hot water.

disqualification \ A feature that makes a chicken ineligible for receiv-ing an award at a show. Disqualifications include any indication of disease, evi-dence of faking, a lack of appropriate characteristics for the breed, and a serious defect or deformity that is likely to be hereditary. Common deformities include crossed beak, side sprigs, humped back, crooked feet, crooked breastbone, and wry tail. Some disqualifications pertain to a specific breed or variety, so a disquali-fication for one breed may be a desirable feature for another.

distress call \ The sound a chicken makes when pecked or caught. A chicken that's suddenly pecked by another chicken lets out a moderately loud startled call that may be either shrill or barely audi-ble, depending on the pecked chicken's temperament, position in the peck order, and how hard it's been pecked. A chicken that has been captured and is being car-ried away makes loud, long, repeated sounds of distress intended to frighten the aggressor into letting go and also to warn other chickens of immediate danger. The other chickens may run and hide, although a courageous cock, or occasionally a hen, may try to rescue the distressed bird by attacking the person or animal carrying it away.

dominant trait \ A trait that

appears when one allele of a gene has greater influence over a variant allele of the same gene when both alleles are present. Examples of dominant traits include crest, five toes, feathered legs, frizzledness, and side sprigs.

Dominecker/Dominicker \

Slang for Dominique, often erroneously applied to the Barred Plymouth Rock, which was developed from the Dominique.

Dominique \ A breed of unknown

origin, brought to the United States by early settlers and kept as a dual-purpose farmstead breed that does well in cold weather. The breed name may derive from early chickens brought from the French colony of Saint-Domingue (now Haiti). The Dominique has a rose comb with a short upward turning spike, may be large or bantam, and comes in one color — irregular barring, or cuckoo. The color is similar to that of the more regularly barred Plymouth Rock, which was developed from the Dominique and with which the Dominique is often confused, but the two breeds may be easily distinguished by comb. Like the Barred Rock, the Dominique is an autosexing breed; at hatch Dominique cockerels are a shade or two lighter than pullets and may remain so into maturity. The hens are good layers of brown-shell eggs and are good broodies.

door keeper \

See: automatic door closer

Dorking \ A breed devel-

oped in the area around Dorking, England, from chickens originally introduced by Roman legions in the first century. The Dorking is a large, docile chicken with short legs and five toes. Dorkings come in two comb varieties — large single comb (suitable for warm climates) and rose comb (more suitable for northern areas) — as well as a few color varieties, of which silver gray is the most common. The hens lay eggs with slightly tinted shells and brood easily.

Dominique hen

Dorking hen

double laced \ A color pattern in which the hen's feathers have two concentric black lacings, as may be found in Barnevelders, dark Cornish, and Wyandottes.

double laced

double laced blue \ A color pattern in which the hen's feathers have two concentric blue lacings, as may be found in Barnevelders.

double mating \ Maintaining two distinct breeding lines, one for producing exhibition-quality hens (called the pullet line) and the other for producing exhibition-quality cocks (the cockerel line). Cocks produced by the pullet line, known as pullet-bred males, may be used to continue the pullet line but are generally not of exhibition quality; likewise, hens produced by the cockerel line, known as cockerel-bred hens, may be used to continue the cockerel line but are generally not of exhibition quality.

To get good show cocks, match your best male to females that tend toward cock color or type. To get good show females, mate your best females to a cock with hen color or type. Exhibit only the top cocks from your cockerel line and top hens from your pullet line.

Double mating is generally used for breeds or varieties in which the color markings of hens and cocks tend to differ, such as Barred Plymouth Rocks, penciled Hamburgs, and dark Brahmas. It may also be used to control type by, for example, selecting a pullet-bred Leghorn cock with a small comb and otherwise feminine head, to offset the Leghorn hens' tendency to develop excessively large combs.

double yolker

double yolker \ An egg containing two yolks, generally as a result of ovulation occurring too rapidly. Double yolkers may be laid by a pullet whose production cycle is not yet well synchronized and are occasionally laid by heavy-breed hens, often as an inherited trait. Sometimes an egg contains more than two yolks. The greatest number of yolks on record is nine in one egg.

down \ The soft furlike fluff covering a newly hatched chick. \ The fluffy part of any feather, below the web where the barbs don't hook together.
[Also called: fluff]
\ An undesirable cluster of fluff on the shanks or feet of a clean-leg breed or variety.

down sheaths \ Small, quill-like structures encasing each individual bit of down on a newly hatched chick. As the chick dries, the down sheaths rupture and fall away, creating a lot of fluff debris.

draft \ An undesirable current of air. To detect a draft in your chicken coop, hold out a strip of survey tape or tissue paper as you move around the coop, pausing quietly at various spots for a few seconds to see if the ribbon moves while you're standing still. Raise your arm to check for overhead drafts, especially at roost height, and squat to check for drafts at chicken height. If the tape moves while you are still, you have a draft, and the direction the tape blows shows you where the draft is coming from. A draft may be a welcome cooling breeze in hot weather, but in cold weather it can cause your birds to chill and suffer frostbite.   *See also: ventilation*

draggy hatch \ Incubated eggs that take more than 24 hours for all the chicks to emerge from their shells. Draggy hatch may be caused by improper incubation temperature or humidity or may result from storing eggs over a long period of time while collecting enough to fill the incubator — the eggs stored longer will take longer to hatch. Another cause is combining eggs of various sizes in one setting. Larger eggs take longer to hatch than smaller eggs; when eggs from bantams and large fowl, or from light and heavy breeds, are incubated together, the bantam eggs or those from the lighter breeds usually hatch first. When a hatch is draggy, remove the dried chicks every 6 to 8 hours so the incubator's airflow won't dehydrate them. Work quickly, since opening the incubator causes the temperature and humidity to drop, reducing the percentage of remaining eggs that will hatch.

draw \ To remove the internal organs of a chicken being butchered for meat. *[Also called: eviscerate]*

drench \ To give liquid medication by mouth. \ A liquid medication put into a chicken's mouth or down its throat.

drinker \ *See: waterer; waterer, automatic; waterer, bell; waterer, chick; waterer, gravity flow*

drinking station \ *See: watering station*

droopy wing

droopy wing \ A wing carried so loosely the secondary feathers fail to cover part of the primary feathers. Droopy wing is characteristic of some breeds including Japanese and Serama. A wing that droops gradually as a chick grows may indicate a structural defect, such as weak muscles of the wing or shoulder. A wing that suddenly starts drooping, or appears in a number of birds, is likely a sign of disease.

droppings \ See box on page 91.

droppings boards \ Wooden slats, generally consisting of 1- by 2-inch (2.5 to 5 cm) lumber placed on edge for rigidity, with 1-inch (2.5 cm) gaps between them, covering a droppings pit.

droppings color \ Normally brown, grayish brown, or greenish brown capped with white or tan urinates (the latter deposited primarily overnight). Softer cecal droppings are usually mustard yellow, chocolate brown, or greenish brown. Bright yellow, bright blue, or bright green droppings may indicate a disease or a toxic reaction to something eaten. Bloody droppings are a sign of disease, typically coccidiosis, although a slight reddish tinge may simply be a bit of sloughed-off intestinal tissue. Watery, off-color droppings during the heat of summer usually indicate the chickens aren't getting enough to drink.

droppings pit

droppings pit \ An area beneath roosts that is covered with wire mesh or droppings boards, beneath which night-deposited manure accumulates where chickens can't scratch or peck in it. Not only do the chickens remain healthier, but the droppings are easier to remove because they don't get trampled and packed down. Bedding spread at least 2 inches (5 cm) thick under the droppings boards will absorb moisture from manure and make scoop-up easier. *[Also called: manure box]*

drumette

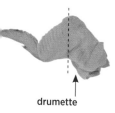

drumstick

droppings tray \ A flat, shallow container placed underneath a cage or beneath a roost to collect droppings for quick disposal. A thin layer of litter, an opened-out feed sack, or some newspaper on which manure accumulates will make the tray easy to clean.

dropsy \ *See: ascites*

drug residue \ The amount of a medication remaining in a chicken's system after most of the drug has metabolized. Drug residue is an important consideration with respect to chicken meat or eggs fed to humans, as it can contribute to human resistance to prescription drugs. Medications in rations fed to industrially produced chickens include low levels of antibiotics to improve feed conversion. However, antibiotics are not included in prepared rations sold for backyard use. The primary drug residues in backyard chickens (unless they are otherwise medicated) come from dewormers and coccidiostats, both of which have established withdrawal periods as specified on the labels.

drumette \ The upper, meaty, section of the wing.

drumstick \ The lower thigh.

dry picking \ A method of plucking a recently butchered chicken. Dry picking is easy if a chicken is freshly killed and still warm, especially if it has been debrained, but becomes more difficult as the chicken starts to cool. It involves quickly stripping away feathers while the chicken is hung for bleeding. Anyone sensitive to chicken feathers or dander should wear a dust mask while dry picking.

dual-purpose breed \ A breed raised for the production of both eggs and meat. Dual-purpose chickens don't lay as well as laying breeds and don't

D

DROPPINGS

Chickens expel two different kinds of excrement. Frequent intestinal droppings consist of firm, grayish brown feces capped with white urine salts. Approximately every tenth dropping is a cecal dropping.

Chicken droppings are made up largely of undigested feed combined with intestinal bacteria, digestive juices, mineral by-products from metabolic processes, and water. The urine salts, or urates, consist mostly of nitrogenous waste and water. Moisture makes up about 80 percent of the total weight of fresh droppings, and its evaporation contributes to henhouse humidity, not to mention odor.

To calculate the total amount of droppings chickens deposit, keep track of feed input and figure about 1.15 pounds of fresh droppings per pound of feed consumed. As droppings age and moisture evaporates, the weight goes down. Fully dry droppings have lost up to 80 percent of their fresh weight. Aged, but not fully dry, droppings on average weigh about one-third as much as fresh droppings, not including any bedding mixed in. A large percentage of droppings are deposited beneath roosts overnight, and the remainder is spread around outdoors during the day.

See also: manure

APPROXIMATE DROPPINGS OUTPUT PER CHICKEN

	Fresh	Aged
HYBRID BROILER (TOTAL)*	11.5 lb (5 kg)	4 lb (2 kg)
NONHYBRID BROILER (TOTAL)*	23 lb (10.5 kg)	8 lb (3.5 kg)
BANTAM (ANNUAL)	30 lb (13.5 kg)	10 lb (4.5 kg)
LIGHT BREED (ANNUAL)	120 lb (54.5 kg)	40 lb (18 kg)
MIDWEIGHT BREED (ANNUAL)	180 lb (82 kg)	60 lb (27 kg)
HEAVY BREED (ANNUAL)	240 lb (109 kg)	80 lb (36 kg)

*Average broiler's total output from hatch to slaughter

duck foot

(normal foot)

grow as efficiently as meat breeds, but lay better than meat breeds and grow faster and larger than laying breeds. Dual-purpose breeds are the quintessential farmstead chickens. Most breeds in the American and English classifications are dual-purpose, although many others, including some breeds typically considered ornamental, are equally suitable for this purpose.

A few hybrids have been developed as efficient dual-purpose birds, including black sex-links and red sex-links. However, if the purpose for keeping a dual-purpose flock is self-sufficiency, which includes hatching your own future replacements, hybrids are not the way to go, since they do not breed true. *[Also called: utility breed]*

dub To surgically trim the comb of a cockerel, usually accompanied by cropping the wattles and ear lobes, thereby leaving the bird's head and face smooth. Dubbing is generally done for one of the following purposes:

- To exhibit mature Modern or Old English Game cocks, which must be dubbed and cropped to qualify for most shows. Exhibitors who don't believe in dubbing and cropping, or who live in a state where doing so is illegal, show only cockerels under six months of age, as well as pullets and hens, and reserve intact cocks for breeding.
- To prevent or treat frostbite in a large-comb breed. *See: frostbite*
- To minimize blood loss in fighting cocks, which otherwise tear into each others' comb or wattles. In some states owning dubbed cocks is considered proof the cocks are kept for fighting, causing legal problems for exhibitors of Modern or Old English Game cocks.

**duck foot ** An undesirable position of the hind toe, which does not point straight back and lie firmly on the ground to properly balance the bird, but instead points toward the front along with the other toes, making the chicken's foot look more like a duck's.

dub and crop

before

after

duplex comb \ A comb that divides into two distinct parts. Duplex comb styles include the buttercup comb and the V- or horn comb.

dusky \ Dark shading overlaying a lighter color, as in the dusky yellow shanks and toes of a brown Cochin, or a black or gray Japanese bantam, and the dusky horn beak of a silver or golden Polish.
[Also called: swarthy]

dust bath \ A shallow bowl created by chickens in soft soil, loose sand, or deep bedding where they thrash around and cover themselves with dust. This activity is so natural that chickens, and baby chicks, engage in dusting behavior even when no dust hole is available. While bathing, chickens work dust through their feathers by flapping their wings and kicking their legs. When a chicken is done dusting, it typically stands up and shakes itself off, then preens.

To enhance parasite control, poultry keepers sometimes add wood ashes, diatomaceous earth, or sulfur garden powder to the dust hole. But chickens are highly susceptible to respiratory problems, and breathing in these foreign materials can make matters worse. On the other hand, for chickens that are seriously infested, the benefit may outweigh the danger of temporarily adding such materials.

dust bath

D

Dutch Bantam \ A true bantam breed named after the seaport town of Bantam on the island of Java in Indonesia, brought to the Netherlands in the 1600s by spice ships. The word bantam has since become synonymous with small chickens. The original color was the same as that of red jungle fowl, leading many authorities to speculate the Dutch Bantam may be among the first domesticated breeds. Dutch Bantams are perky, active birds and proficient fliers.

They have a single comb and come in several color varieties. The hens lay eggs with pale brown shells and brood easily.

dwarf egg \ A yolkless egg. It is sometimes called a dwarf egg because it is considerably smaller than most eggs laid by the same hen. *See also: yolkless egg*

Dutch Bantam
hen

lay blue-shell eggs; and Empordanesa and Pendesenca, which have white earlobes but lay eggs with dark brown shells.

ear ring \ An ear tuft.

ear tuft \ A clump of feathers sprouting from a tiny finger-like appendage (the peduncle) made of cartilage and protruding from the side of the neck just below the ear, as is characteristic of Araucanas. Ideal ear tufts have a good length, and match in size, shape, and location. They extend in well-defined ringlets on a cock or well-defined spoon-like curves on a hen, although tufts may vary in shape from spiral to spherical to fan-shaped to bunched-up like a rose. Tufts also vary in size, small ones generally growing closer to the ears, larger ones lower down the neck. Some Araucanas have an ear tuft on only one side, some have more than one tuft on a side, and some have none at all.

   The tufted trait is carried by a lethal gene that affects the structure of the ear canal. A chick inheriting a tufted gene from both parents has deformed ears and mouth, and dies before hatching, typically during the final days of incubation. Even a single tufted gene can interfere with good hatchability, making Araucanas with ideal conformation difficult to breed.
[Also called: ear ring; ear whisker]

ear whisker \ An ear tuft.

ear

earlobe

ear tuft

ear \ The small opening on each side of the head, below and behind the eyes, and covered by a clump of tiny, stiff feathers. Unlike a dog, cat, or human, a chicken has no external ear part, so its ears might be considered more like ear holes. Although its ear structure is not as complex as that of a human, the chicken can hear pretty well.

earlobe \ The patch of bare skin below the ear, varying in size and shape according to the breed. In all breeds the texture should be fine and soft, the surface smooth and free of folds or wrinkles, and the two earlobes matching in size and shape.

   Earlobes are usually either red or white, but in some breeds they are turquoise blue (as in Silkies and some Sebrights) or purple (Sebrights and Sumatras). In general, hens with white earlobes lay white-shell eggs, and hens with red earlobes lay brown eggs. Exceptions are Crevecoeur, Dorking, and Redcap, which have red earlobes but lay white-shell eggs; Araucana and Ameraucana, which have red earlobes but

Easter Eggers (EE) \ Chickens with Araucana or Ameraucana blood but lacking the standard conformation or color varieties of either breed. They are called Easter Egger chickens because they produce eggs with a variety of pretty shell colors ranging in shades from green to greenish blue, from brown to pinkish, and from white to pale yellow or gray.

E. coli \ *See: Escherichia coli*

edging \ A narrow band or lacing of contrasting color around all or part of a feather, as is characteristic of the hackle feathers of Columbian varieties.

edged feather

egg \ An oval object laid by a hen that contains nearly all the nutrients necessary for life, lacking only vitamin C. An egg is made up of several layers of albumen (egg white), several layers of yolk, and several layers of shell. A large egg is approximately 31 percent yolk, 58 percent white, and 11 percent shell. Aside from the shell, an egg has about 75 percent water, and the remainder is about 12 percent protein and 12 percent fat. Eggs to be used for hatching, or hatching eggs, must be fertile; eggs intended for eating, or table eggs, may or may not be fertile. **See page 54 for illustration of egg anatomy.**

egg, dirty \ Eggs are clean when laid, and if the nests are properly designed and managed, the eggs should be clean when collected. Occasionally, a really nasty egg is found on the coop floor or in a nest where a chicken has roosted on the edge and soiled the litter.

Such an egg is covered with bacteria and therefore not safe to eat or hatch — discard it.

Soiled eggs within the nest may result from layers tracking mud or muck on their feet. To keep eggs from getting dirty, eliminate the source of mud — most often a muddy entry or damp ground around a leaky drinker. Eggs in nests located on or near the floor are more likely to get dirty than eggs in nests raised above the floor, especially if layers must reach the nests by hopping up on a rail or series of rails. Since chickens like to roost on these rails, make sure none is so close to a nest that a roosting chicken would be able to fill the nest with droppings. For most breeds a rail no closer than 8 inches (20 cm) from the nest's edge should work. *See also: egg cleaning; egg washing*

egg, hatching \ *See: hatching egg; incubate; incubation humidity; incubation period; incubation temperature*

egg, infertile \ Eggs laid by a flock of hens with no rooster present will not be fertile. They are therefore incapable of hatching but are fine for eating. *See also: fertility, low; fertility, weak*

egg, wormy \ An egg containing one or more worms. Wormy eggs are rare, as they are laid only by hens that are seriously infested with internal parasites.

egg basket \ A container for collecting and transporting eggs, which should be of adequate size for the number of eggs involved so they don't get cracked by rolling around and banging together. Baskets designed specifically for collecting eggs are available from poultry suppliers, although nice wire baskets may be found at a hobby shop or general store.

egg binding, egg bound \

A condition in which an egg gets stuck just inside the vent, usually because the egg is too large, the pullet is fat or unhealthy, or the pullet's body isn't fully mature when she starts laying. Egg binding can be an extremely serious condition, especially if the pullet goes into shock.

First make sure the pullet is truly egg bound. If she strains to release an egg, and the end of the egg may be seen near the opening, she's egg bound. If the egg is not visible, lubricate a finger with K-Y Jelly or other water-based lubricant and gently insert it into the vent until you feel the hard shell with the end of your finger. Do not attempt to stretch the vent, which could tear delicate tissue.

Once you are certain the pullet is egg bound, lubricate the vent and egg with K-Y Jelly and/or gently squirt in warm (not hot) saline-solution wound wash or warm soapy water. Gently insert your lubricated finger to help maneuver the egg, while with your other hand push gently against the abdomen and try to work the egg out. Be careful not to break the egg, which can cause internal injury.

When an egg won't slide out easily, warming the vent area may relax the muscles enough to release the egg. Moisten an old towel, warm it in the microwave (make sure it's not hot), and apply it to her bottom. Reheat the towel as needed to keep it warm, or use two towels and warm them alternately, to maintain moist heat. If you don't have a microwave oven handy, put warm, not hot, water in a bucket or basin and stand the pullet in it with the water reaching just above her vent. After warming the pullet's bottom for about 15 minutes, give her a rest, and if she doesn't soon release the egg, try again.

egg carton \ An enclosed container

for storing eggs to keep them fresh for eating. A traditional egg carton holds a dozen or a dozen and a half eggs, although smaller cartons are available that hold only four or six eggs. A carton with a solid cover keeps eggs fresh longer than a carton with part of the top cut away to make the eggs more visible. Egg cartons may be recycled sparingly, but eventually, they become unsanitary. A washable, reusable carton is a better option for storing eggs from backyard chickens.

EGG DEVELOPMENT

A hen's ovary contains a large number of undeveloped egg yolks, which mature one by one. Approximately every 25 hours, one yolk matures and is released into the funnel of the oviduct, a process called ovulation, which usually occurs within 1 hour after the hen laid her previous egg. During the yolk's journey through the oviduct, it is fertilized (if sperm are present), encased in various layers of egg white, wrapped in protective membranes, sealed within a shell, and finally enveloped in a fast-drying fluid coating called the bloom, or cuticle.

Throughout its passage through the 25-inch-long oviduct, the egg leads with its pointed end. Just before it is laid, the egg rotates so the blunt end comes out first, which keeps the egg from cracking when it plops into the nest. Since the whole process takes about 25 hours, a hen lays her egg about an hour later each day. And since a hen's reproductive system slows down during the night, eventually, she'll skip a day altogether and start a new multiple-day laying cycle the following morning.

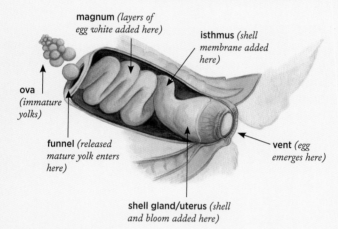

magnum *(layers of egg white added here)*

isthmus *(shell membrane added here)*

ova *(immature yolks)*

funnel *(released mature yolk enters here)*

vent *(egg emerges here)*

shell gland/uterus *(shell and bloom added here)*

egg cleaning \ A slightly dirty egg may be brushed off or rubbed with a sanding sponge or nylon scouring pad. An egg that needs more than light cleaning should be either washed or, preferably, discarded. *See also: egg washing*

egg collection \ The sooner freshly laid eggs are collected from the nest and properly stored, the longer they'll stay fresh. Hens generally lay in the morning, but some lay a little later each day until about midafternoon. By knowing what time of day your hens lay, you can coordinate egg collection to coincide. Otherwise collect eggs often, preferably two or three times a day, to keep them from getting dirty or cracked, spoiled in hot weather, or frozen in cold weather.

egg eating \ A form of cannibalism that usually starts when eggs get broken in the nest, either because too few nests cause hens to crowd together or because a nutritional deficiency causes shells to be thin. Once chickens find out how good eggs taste, they break them on purpose to eat them. The only way to stop egg eating and keep it from spreading is to remove the culprit early. Identify instigators by checking for egg yolk smeared on beaks or by catching the eaters in the act. When eggs disappear entirely, or shells that have been cleaned of their gooey contents are found

E

in or near the nest, the problem may not be an egg-eating chicken but an egg-loving predator.

egg fertility \ *See: fertility*

egg flavor \ Off-flavor eggs may result from something a hen ate or from environmental odors. Hens that eat onions, garlic, fruit peelings, fish meal, fish oil, or excessive amounts of flaxseed may lay eggs with an undesirable flavor. Eggs can also absorb odors that translate into unpleasant flavors if they're stored near kerosene, carbolic acid, mold, must, fruits, and vegetables.

egg floating \ Floating an egg in plain water to gauge the size of its air cell. A fresh egg settles to the bottom of the container and rests horizontally. The larger air cell of a one-week-old egg causes the big end to rise up slightly from the container bottom. An egg that's two to three weeks old stands vertically at the bottom of the container, big end upward. When the air cell grows large enough to make the egg buoyant, the egg will float. A floating egg is quite old but not necessarily unsafe to eat.

egg freshness \ You may on occasion find an egg, or a cache of eggs, and not know how long ago they were laid. You can estimate an egg's age by candling it, floating it, checking its odor, or breaking it open to examine the interior quality.

The albumen of a fresh egg contains carbon dioxide that makes the white look cloudy. As an egg ages, the gas escapes and the albumen looks clear or transparent. A fresh egg's albumen is firm and holds the yolk up high. A stale egg has watery albumen that spreads out thinly around the yolk. As an egg ages, water migrates from the albumen to the yolk, stretching and weakening the yolk membrane. The older the egg, the greater the likelihood that its yolk will break. But even the freshest egg occasionally has a watery white or an easily broken yolk.

If your flock includes one or more roosters, crack the found egg into a separate container before using it in cooking, to ascertain it doesn't contain a developing embryo because of partial incubation by a hen. Any egg that has a mottled yolk or otherwise doesn't look right should be discarded.

egg freezing \ During winter weather an egg left in the nest may freeze, causing the contents to expand and put pressure against the shell until it cracks. A cracked egg is not safe for eating and should be discarded. Freezing of eggs in the nest may be prevented by frequent egg collection during cold weather.

egg grading \ Sorting eggs according to exterior and interior qualities into three grades established by the United

States Department of Agriculture. For all grades the shell must be intact. Grades AA and A are nearly identical. Grade AA eggs are slightly fresher. Grade A eggs are most commonly found in grocery stores. Grade B eggs have minor defects and are not sold to consumers. In the industry they are used in prepared egg products; homegrown Grade B eggs are best used for recipes in which eggs are stirred and cooked. Nutritionally, all grades are the same.   *See also: air cell gauge*

egg laying \ At a hen's cloaca, just inside the vent, the reproductive and excretory tracts meet, which means a chicken lays eggs and poops out of the same opening — but not at the same time. As an egg is pushed out into the world, the bottom end of the oviduct turns inside out, wrapping around the egg and pressing shut the intestinal opening, so the egg emerges clean.

eggmobile \ *See: chicken mobile home*

egg odor \ The foul-smelling odor of a rotting egg, caused by hydrogen sulfide, otherwise known as rotten-egg gas. Any egg with an off-odor, whether raw or cooked, should not be eaten.

egg quality \ The appearance and consistency of an egg's contents, as determined by breaking the egg into a dish for examination. As an egg ages, both its white and yolk deteriorate. Their quality may not have been all that great to start with, depending on the hen's age and health, the use of medications, the weather, and hereditary factors. The

better an egg's starting quality, the better it keeps.

The occasional egg is abnormal due to an accidental occurrence, the hen's hereditary tendencies, or environmental or management factors. Some egg abnormalities are little more than nonrecurring curiosities, and others may require corrective action. Such abnormalities may be detected by candling and by inspecting broken-out eggs. Trapnesting is a way to identify hens that habitually lay poor-quality eggs.   *See also: egg freshness*

egg settling \ Giving shipped hatching eggs a rest for a few hours before placing them in the incubator. Settling has two purposes: to give the eggs a rest after being jostled during transport, and to give them time to adjust to a uniform temperature before being placed in the incubator.

egg sexing \ Determining the sex of a fertilized egg or a developing embryo by examining the outside of the shell. Although egg sexing based on shell shape has long been practiced in Southeast Asia, if the method were reasonably reliable, commercial poultry producers would not continue to spend time and money looking for ways to sex eggs before they hatch. To date the best methods industry has developed are tedious and expensive. One involves magnetic resonance imaging to determine if a developing embryo has male or female organs. Another involves using a needle to remove a sample from an incubating egg to find out if it contains estrogen compounds, indicating the embryo is female.

E

egg shape \ Most chicken eggs have a rounded or blunt end and a more pointed end, although some eggs are nearly round, while others are more elongated. An egg's shape is established in the part of the oviduct called the isthmus, where the yolk and white are wrapped in shell membranes. An egg that for some reason gets laid after being enclosed in membranes but before the shell is added has the same shape it would if it had a shell. Each hen lays eggs of a characteristic shape, so you can usually identify which hen laid a particular egg by its shape.

The curved surface is designed to distribute pressure evenly, provided the pressure is applied at the ends of the egg, not at the middle. The middle of a shell must be weak enough to allow an emerging chick to peck all around and break out of an incubated egg. By contrast the ends of an egg must be quite strong so a newly laid egg won't crack when it plops into a nest, blunt end down.

egg shape

eggshell \ *See: shell*

egg size \ The United States Department of Agriculture sets standards for eggs in six sizes, according to minimum weight per dozen eggs, to account for slight variations from one egg to the next. On a per-egg basis, the smallest size is peewee, with an average weight of 1.25 ounces (35 g), and the largest is jumbo, weighing about 2.5 ounces (70 g). Periodically, some poultry keeper will report finding a chicken egg weighing 6 ounces (170 g) or more. The heaviest chicken egg on record weighed 1 pound (0.5 kg).

A pullet starts out laying small eggs, but as time goes by her eggs reach normal

EGG SIZES

Size	Weight per Dozen		Average Weight per Egg		% of Medium Egg
	OUNCES	GRAMS	OUNCES	GRAMS	
PEEWEE	15	425	1.25	35	71%
SMALL	18	510	1.50	43	86%
MEDIUM	21	595	1.75	50	100%
LARGE	24	680	2.00	57	114%
EXTRA LARGE	27	765	2.25	64	129%
JUMBO	30	850	2.50	70	143%

size for her breed. As a hen ages, after each annual molt her eggs typically get a little bigger. Egg size is also influenced by a hen's weight. Pullets that are underweight when they start laying will continue to lay smaller eggs than birds that mature properly. A hen of any age and weight may temporarily lay small eggs if she is suffering from stress induced by heat, crowding, or poor nutrition, including inadequate protein or salt.

Among same-age hens of a given breed or strain, the majority of eggs should be of one size, with only an occasional egg ranging just above or below the majority.

egg storage \ TABLE EGGS should

be stored in an egg carton, oriented with the large ends upward to keep the yolks nicely centered. Refrigeration is usually the quickest, most convenient way to store table eggs, but the worst place to store them is on an egg rack built into the refrigerator door, where they get jostled every time the door is opened. And if the rack lacks a cover, the eggs will be exposed to lost moisture and blasts of warm air whenever the door is opened.

Storing closed cartons on a lower shelf keeps eggs cooler, thus fresher longer, and reduces evaporation through the shell. On the lowest shelf, where the temperature is coolest, eggs in a closed carton will keep for up to five weeks. The biggest problem with a refrigerator is its low humidity, especially in a self-defrosting (frost-free) model. By wrapping cartons in plastic bags to prevent moisture loss (as well as absorption of flavors from other foods), you can safely refrigerate eggs for two months.

Once whole eggs have been refrigerated, they need to stay that way. A cold egg left at room temperature sweats, encouraging the growth of bacteria. Never leave a refrigerated egg on the counter for more than two hours.

HATCHING EGGS should never be refrigerated. Store them away from sunlight in a cool, relatively dry place, preferably with a temperature of 55°F (13°C) and humidity low enough to prevent moisture from condensing on the shells, which would attract molds and also encourage any bacteria already on the shell to multiply. Store hatching eggs in clean cartons with their large ends up to prevent the yolk from sticking to the inside of the shell at the pointed end.

The less moisture that evaporates from eggs during storage, the better chance they have of hatching. Small eggs laid by bantams and jungle fowl have a relatively large surface-to-volume ratio and therefore evaporate more quickly than larger eggs. Late-summer eggs of any size have thinner shells because the hens have been calling on their calcium reserves all summer. These shells allow more rapid evaporation than occurs in early-season eggs.

The longer an egg is stored, the longer it will take to hatch, and its ability to hatch gradually decreases. Hatching eggs properly stored for up to 6 days generally retain good hatchability, but for each day thereafter hatchability suffers by approximately 1 percent. Evaporation may be minimized for eggs stored longer than 6 days by wrapping cartons in plastic bags. Yolk position may be maintained by tilting cartons from one side to the other each day. Eggs stored in this manner retain decent hatchability for up to 10 days.

egg tooth \ A small, sharp temporary cap at the tip of a newly hatched chick's upper beak, which falls off a few days after the chick hatches. As hatching time approaches, and the chick can no longer get enough oxygen through the shell pores, it uses its egg tooth to break into the air cell at the large end of the egg. There it finds sufficient oxygen to give it time to break out of the shell. A pipping muscle at the back of the tiny bird's neck gives it enough strength to use the egg tooth to pip through the shell's outer membrane and then through the shell itself to make its escape.

egg tray \ A tray made of plastic or paper pulp, usually designed to hold 30 eggs. *See also: flat*

egg turning \ The rotating of eggs during incubation. By fidgeting in her nest and adjusting eggs with her beak, a setting hen periodically turns the eggs beneath her, in the process keeping yolks centered so they don't stick to the shell lining. To imitate the hen's activities, eggs in an incubator must be periodically turned. Some incubators have automatic turners. Without a turner, eggs need to be manually turned at least three times a day. To make sure each egg is turned, place an X on one side and an O on the opposite side with a grease pencil or china marker.

Eggs aren't turned end to end, but side to side. Throughout incubation the pointed end of the egg should never be oriented above the blunt end. Otherwise, fewer chicks will hatch, and the ones that do hatch will be of lower quality.

Signs of improper turning are early embryo deaths due to stuck yolks and full-term chicks that fail to pip. Eggs

egg tooth

need not be turned after the 14th day of incubation and should not be turned during the last 3 days before the hatch (at 21 days), when chicks need time to get oriented and begin breaking out of the shells.

egg washing \ Cleaning a moderately soiled egg with water. Egg washing also rinses away the bloom that seals the shell pores and keeps out bacteria. Washing is therefore not an ideal routine practice but might be done to clean a shell smeared with egg white or yolk from a broken egg or to clean a valuable but soiled hatching egg.

First rinse off the egg in water that's slightly warmer than the egg; water that's cooler than the egg can cause bacteria to be drawn through the shell. Sanitize the cleaned egg by dipping it for 30 seconds in a solution of 1 teaspoon of chlorine bleach and 1 quart of water warmed to 101°F (38°C). Wipe the shell dry with a clean paper towel or soft cloth before placing it in a storage carton. When washed eggs will be used for eating, their shelf life may be prolonged by rubbing the shells with clean vegetable oil to replace the natural bloom; do not oil hatching eggs.

egg white \ Approximately 40 different proteins, combined with water, that make up an egg's transparent portion, which turns white when heated. The egg white accounts for about 67 percent of the egg's liquid weight and consists of these four layers:

OUTER THIN. A narrow layer of watery white lying next to the shell membrane.

It repels bacteria by virtue of its alkalinity and its lack of nutrients needed by bacteria for growth.

OUTER THICK, OR FIRM, WHITE. A gel that makes up the greater portion of the egg white.

INNER THIN. The watery white surrounding the inner thick layer and separating it from the outer thick layer.

INNER THICK, OR CHALAZIFEROUS. The layer that surrounds the yolk. It cushions the yolk and contains defenses against bacteria. As an egg forms in the oviduct, it rotates, causing the ends of this layer to twist together to form a fibrous cord, or chalaza, on two sides of the yolk. These cords anchor the chalaziferous layer, protect the yolk by centering it within the shell, and keep the germinal disc oriented upward.

The ratio of thick to thin white varies, depending on the individual hen and the age of the egg. The older an egg gets, the less thick white and the more thin white it has. This thinning allows

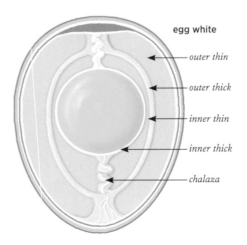

egg white

outer thin

outer thick

inner thin

inner thick

chalaza

the yolk to move around more within the layers of white and also accounts for why, the older an egg gets, the more it spreads when cracked into a frying pan.
[Also called: albumen (from the Latin word albus, meaning white)]

Egyptian Fayoumi \ *See: Fayoumi*

Eimeria infection \ *See: coccidiosis*

electric net \ An all-electric net fence designed for confining poultry, often combined with a battery-powered controller and used as a lightweight, movable fence for range rotation.

electrolytes \ Natural salts and other minerals that help regulate body processes, maintain hydration, and preserve the body's acid/base balance. They are called electrolytes (from the Greek words *electro*, meaning released, and *lytos*, meaning dissolvable) because when dissolved in water they split into electrically charged particles (ions) that transmit electrical impulses.

Dehydration caused by diarrhea, worms, and other conditions causes these minerals — primarily calcium, chloride, magnesium, phosphorus, potassium, and sodium — to be depleted from body fluids. Any time a chicken has loose droppings, or otherwise suffers from dehydration, adding an electrolyte supplement to its drinking water will help its body replace and retain fluids. Electrolytes given to a healthy chicken, however, can cause an imbalance opposite to the desired effect.

Electrolyte mixtures designed for chickens are available in several brands. Gatorade, diluted with an equal amount of water, works in a pinch. A homemade electrolyte solution may also be mixed up from ingredients commonly found in the kitchen. Offer an electrolyte solution in place of drinking water four to six hours a day for no more than a week; at other times replace the solution with fresh drinking water. If chickens are suffering from coccidiosis, an electrolyte solution is *not* a substitute for a coccidiostat.
[Also called: lytes]

HOMEMADE ELECTROLYTE SOLUTION

Ingredient	Source	Amount
POTASSIUM CHLORIDE	Salt substitute	½ teaspoon
SODIUM BICARBONATE	Baking soda	1 teaspoon
SODIUM CHLORIDE	Table salt	1 teaspoon
SUCROSE	Sugar	1 tablespoon
WATER		1 gallon

Administer this solution to a dehydrated chicken in place of drinking water for four to six hours per day for a week, offering fresh water for the remainder of each day.

Electronet \ A brand of electric net fence.

embryo \ A fertilized egg at any stage of development prior to hatching. Developing embryos are a sign the eggs have been partially incubated, which occurs when a hen starts setting but later gives up. It may also occur when eggs accumulate in a nest. While laying subsequent eggs, the hen warms earlier eggs enough for them to start developing, which is especially likely to occur when more than one hen lays in the same nest.

embryo death \ The majority of developing embryos that die during normal incubation do so at two peak times. The first major embryo death loss occurs within a few days of the beginning of incubation, indicating improper egg handling or storage. The second, usually larger death loss occurs just before the hatch and may result from a variety of causes. These include:

- bacterial contamination (dirty eggs placed in the incubator)
- inadequate breeder flock diet
- improper operation of the incubator
- hereditary weakness
- failure to keep the large end of the egg upward during incubation or hatching to enable chicks to break into the air cell

Empordanesa \ A layer breed from the Emporda community in the Principality of Catalonia, Spain. The Empordanesa is closely related to the Penedesenca but is slightly smaller and has a body type similar to that of a Leghorn. It has a carnation comb (the hen's comb may flop to one side), has no bantam counterpart, and comes in a limited number of solid color varieties (but not black), of which white is the most common in North America. Like other Mediterranean laying breeds, the Empordanesa is lean and flighty. The hens are good layers and, despite having white earlobes, produce eggs with dark chocolate-brown shells, sometimes speckled, and do not brood.

enamel white \ The satiny white color typical of the earlobes of white-faced black Spanish and other Mediterranean breeds, as well as of Rosecomb bantams.

endangered breeds \ Chicken breeds having alarmingly low populations, as tracked by three organizations:

- **The American Livestock Breeds Conservancy** periodically conducts a survey to identify endangered old-time production breeds significant to the United States

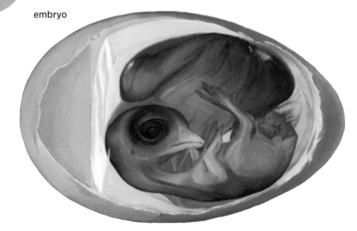
embryo

E

- **Rare Breeds Canada** periodically publishes a conservation list of Canada's rare heritage breeds
- **The Society for the Preservation of Poultry Antiquities** publishes an annual critical list of historical show breeds

See also: rare breeds

endoparasite \ *See: parasite, internal*

energy \ In dietary terms, calories needed to fuel living cells within an animal's body. For chickens, carbohydrates in the form of starchy grains such as corn, oats, and wheat are the main source of energy.

EMBRYO DYEING

Injecting dye into incubated eggs before they hatch can color-identify chicks of different matings. Dye works best on chicks with white or light-colored down and wears off as soon as the chicks grow their first feathers. The colors red, green, and blue show up best. Purple, made by combining red and blue, also works well. Yellow and orange don't show up well at all, especially on chicks with yellow down. Embryos dye best during the 11th to 14th day of incubation. To avoid chilling the eggs, remove no more eggs from the incubator than may be dyed within 30 minutes.

Dyeing requires a 20-gauge, 1-inch (25 mm)-long hypodermic needle with syringe, a sharp sewing needle of the same size or a little bit larger, and a set of food dyes in 2 or 3 percent concentration. About ½ inch (13 mm) from the tip of the pointed end, sanitize an area the size of a quarter using 95 percent rubbing alcohol or povidone-iodine (Betadine is a brand name). Dip the sewing needle into the alcohol or iodine. Cushioning the egg in one hand, make a tiny hole in the center of the sanitized area by pressing against it with the needle, twisting the needle back and forth until it just penetrates the shell and membranes. Take care to make only a tiny hole that does not go deeper than necessary to pierce the inside membrane (no more than ⅛ inch [3 mm]).

Dip the hypodermic needle into alcohol or iodine and fill it with ½ cc of dye. Insert it into the hole so the tip is just beneath the inner shell membrane. Slowly depress the plunger to release the dye without letting it overflow. To avoid inadvertently mixing two colors, use a clean needle when changing dyes.

Seal the hole with a drop of melted wax or a small piece of sheer strip adhesive bandage. Return the eggs to the incubator, and continue incubating as normal. The chicks will hatch having down of the various colors as dyed.

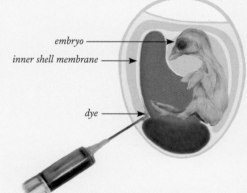

embryo

inner shell membrane

dye

English class \ One of six groupings into which the American Poultry Association organizes large chicken breeds. The breeds in this class, originating primarily in Great Britain and Australia, are Australorp, Cornish, Dorking, Orpington, Redcap, and Sussex.

enteral/enteric disease \ Any disease affecting the intestines and characterized by loose or bloody droppings, weakness, loss of appetite, increased thirst, dehydration, and weight loss in mature birds or slow growth in young birds. Enteric diseases include campylobacteriosis, coccidiosis, colibacillosis (*E. coli* infection), necrotic enteritis, salmonellosis (typhoid, paratyphoid, pullorum), and internal parasites (parasitic worms).

enteritis \ An inflammation of the small intestine.

enterotoxin \ A substance that poisons cells lining the intestines.

entrails \ Internal body organs and glands.
[*Also called: viscera*]

enzootic \ A disease or infectious agent that is continuously present in a specific area or during a particular season, equivalent to an endemic disease among humans.

epizootic \ A disease that is temporarily prevalent and widespread among chickens or other animals, equivalent to an epidemic among humans.

Epsom salts flush \ *See: laxative flush*

Escherichia coli \ Bacteria normally living in the intestines of chickens that, if the birds get out of balance due to stress, nutritional problems, or some other disease, can produce lethal toxins resulting in colibacillosis. These bacteria may infect chickens living in unsanitary or otherwise stressful conditions by entering the digestive or respiratory system. If the bacteria contaminate eggs or chicken meat, they can cause human foodborne illness.
[*Also called: coliform bacteria; E. coli*]

esophagus \ The passageway that moves food from the throat to the stomach.
[*Also called: gullet*]

essential amino acids \ Eight amino acids, collectively, that are not synthesized within a chicken's body but rather by certain plants. A chicken obtains them only by eating a combination of these plants (grains and beans) or by eating animal protein (such as that in bugs or worms) that already contains these amino acids. Any source of protein that furnishes the essential amino acids is considered to be a complete protein.

eversion \ *See: prolapse*

eviscerate \ To remove the internal organs of a chicken being butchered for meat.
[*Also called: draw*]

E

excelsior pads \ Nest pads manufactured from wood shavings and brown paper.

exhibition breed \ A breed kept and shown primarily for aesthetic qualities rather than its ability to lay eggs or efficiently produce meat. Any breed, however — including those developed primarily for egg or meat production — that conforms to the standard description for its breed and variety may be entered into exhibition.
[Also called: show breed]

exotic Newcastle disease \
See: Newcastle disease

Extension \ *See: Cooperative Extension System*

extreme frizzle \ *See: curly*

eye color \ The eye color of chickens varies from breed to breed and may differ for different varieties within a breed. The proper eye color for each breed is part of the standard for that breed. The most common eye color is reddish bay. Brown or dark-brown eyes are also fairly common, found among Australorps, Campines, Javas, Jersey Giants, Langshans, Minorcas, Orpingtons, Spanish, and Sumatras. Pearl is the correct eye color for Aseel, Cornish, Malay, and Shamo chickens. Eyes may also be various shades of red: Phoenix eyes are red, La Fleche bright red, Lakenvelder deep red, and Yokohama orange-red. Modern Game eyes are either red or black, while Old English eyes may be red, black, or brown in addition to

excelsior pad

reddish bay. Araucana eyes may be red or brown in addition to reddish bay.

eyes \ Like most other birds, a chicken has eyes on the sides of its head, giving it a large range of peripheral vision but a small range of binocular vision. Other creatures (including humans) have eyes positioned toward the front so both can focus on an object at the same time. A chicken, by contrast, has a right-eye system and a left-eye system, each with different and complementary capabilities. The right-eye system works best for activities requiring recognition, such as identifying items of food close by on the ground. The left-eye system works best for activities involving depth perception, such as warily watching an approaching hawk.

Compared to most other animals, birds have large eyes in relation to the size of their heads, giving a chicken keen eyesight that is ten times sharper than a human's to help them guard against predators. Chickens have better color vision than most animals, including humans, because their retinas (the

E

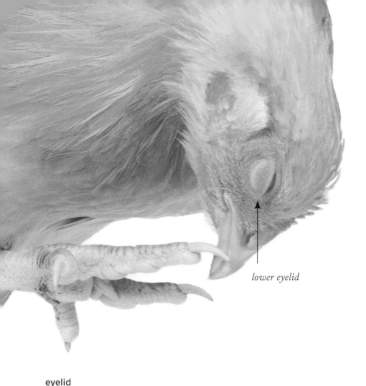

lower eyelid

eyelid

nictitating membrane

light-sensitive area at the back of each eye) are organized primarily for seeing in the daytime, when chickens spend most of their time looking for things to eat. The trade-off of having superior day vision is that they don't see well in the dark.

eyelid \ The protective covering of the eye. A chicken has three eyelids. The upper and lower eyelid are much like a human's, except that a chicken's lower eyelid moves more freely than the upper. The third eyelid, called the nictitating membrane (from the Latin word *nictare*, meaning to blink), lies between the eye and the other two eyelids and has its own lubricating duct, similar to a human's tear duct. The nictitating membrane moves horizontally across the eye from front to rear and is transparent, so the chicken can see even with this eyelid closed. The chicken draws this third eyelid across its eye to clean and moisten the eye and sometimes for protection, such as when a hen's chick tries to peck her eye.

E

F

face \ The soft, smooth skin around and below the eyes on both sides of a chicken's head. Most breeds have a bright red face. Exceptions are the purplish faces of the Sebright, Silkie, and Sumatra and the pure white face of the black Spanish.

fake \ The fraudulent practice of removing or concealing a defect or disqualification from a potential buyer or show judge, thus making the bird look like something it is not. Examples of faking include starching tail feathers to make them stand erect, stitching a wry tail to straighten it, recoloring the beak or earlobes, surgically trimming a faulty comb, and chemically altering the natural color of plumage.

false wing \ Alula. It's called a false wing because it looks like a tiny fake wing growing out of the regular wing. *See also: alula*
See page 11 for illustration.

fancier \ A person involved with the fancy.

fancy \ The appreciation and enjoyment of chickens, typically by owning (and, commonly, breeding and showing) them.

fan-ventilated incubator \ *See: circulated-air incubator; incubator*

fat \ A chicken has some fat in its muscle, and a lot in its skin, but the main measure of fatness is its fat pad.

fat pad \ Abdominal fat. Chickens evolved with the ability to develop a fat pad for use as reserve energy during times when forage is scarce. Most young chickens, especially active pastured birds, have a relatively thin fat pad. An exception is commercial-strain broilers fed for rapid growth; any excess feed they don't convert into muscle (meat) metabolizes into fat. In general an older chicken has a thicker fat pad than a younger chicken, and a hen has a thicker fat pad than a

red face

purple face

cock of the same age. Old hens, especially inactive hens fed too much grain, can accumulate enormous quantities of fat, to the point that the abdominal cavity is virtually filled with it. Signs that a laying hen has an excessively large fat pad include:

- Low rate of lay
- Laying at night
- Poor shell quality
- Frequent multiple yolks
- Prolapse

Faverolle \ A breed developed in France as a dual-purpose farmstead chicken and named after the French village of Faverolle. The Faverolle was bred from so many other breeds that no one can agree on its exact heritage. It has a single comb, a beard and muffs, slight leg feathering, and five toes. The Faverolle may be large or bantam and comes in a few color varieties, of which the original color, salmon, remains the most common. The hens are good layers of eggs with light brown shells and tend toward broodiness.

Fayoumi \ An ancient breed developed in Egypt, where this chicken is widely known; Fayoum, one of the Egyptian governorates, includes both the Fayoum Oasis and the capital city of Fayoum. The Fayoumi is a small chicken with a scrawny, forward-angled neck and an equally scrawny tail that angles forward at almost 90 degrees, giving it an overall appearance that's reminiscent of a roadrunner. These chickens are particularly disease resistant, excellent foragers, and strong fliers. They tend to be somewhat nervous and are extremely vocal when handled. They have a large single comb, come in a single black-and-white color pattern, and have no bantam counterpart. The hens are outstanding layers of small eggs with tinted shells and are not inclined toward broodiness when young but may brood as they age. [*Also called: Egyptian Fayoumi*]

Faverolle hen

Fayoumi hen

F

FEATHER PATTERNS

See also: plumage patterns

spangled

penciled

barred

stippled

double laced

mille fleur

laced

lacing

ground color

edged

striped

FEATHER ANATOMY AND TYPES

A feather is one of many growths coming through a chicken's skin that together make up the plumage covering the bird's body. Feather shape, color, and texture vary with breed and variety, but every chicken has four different kinds of feather.

WEBBED FEATHERS consist of a hollow quill at the base, closest to the skin, and a solid shaft at the top. From the shaft sprout slender barbs bearing barbules, which in turn bear barbicels with interlocking hooks by which they fasten together to form a smooth vane or web. At the lower end of the shaft, just above the quill, the hooks are lacking, causing that portion of the feather to be fluffy rather than smooth. Contour feathers, flight feathers, tail feathers, and coverts are all webbed feathers (except in silkie feathered breeds.)

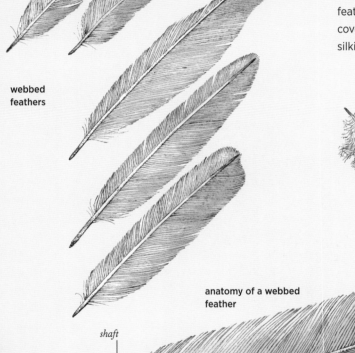

webbed
feathers

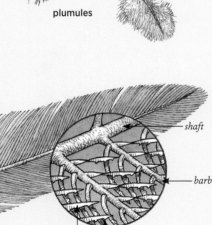

plumules

anatomy of a webbed
feather

shaft

quill

vane/web

shaft

barb

barbule   *barbicel*

F

PLUMULES or down feathers are short and lack barbicels; therefore, they are fluffy rather than smooth. These feathers form a soft undercoat beneath the webbed feathers, trapping air to help the chicken maintain body temperature. *See also: silkie feathered*

FILOPLUMES are small, hairlike feathers attached to nerve endings to monitor the placement of contour feathers, such as when a chicken puffs out its feathers to look fierce or stay warm. The filoplumes appear as hairs remaining on the skin of a freshly butchered and plucked chicken. Old-fashioned breeds have more filoplumes than Cornish-cross broilers, and dark-colored breeds have more visible filoplumes than light-colored breeds.

BRISTLES are the small, stiff feathers with a few barbs at the base, growing around the chicken's eyes, nose, and mouth. They function like a human's nose or ear hairs to protect sensitive organs from dust and insects.

filoplume

bristle

FEATHER POSITIONS

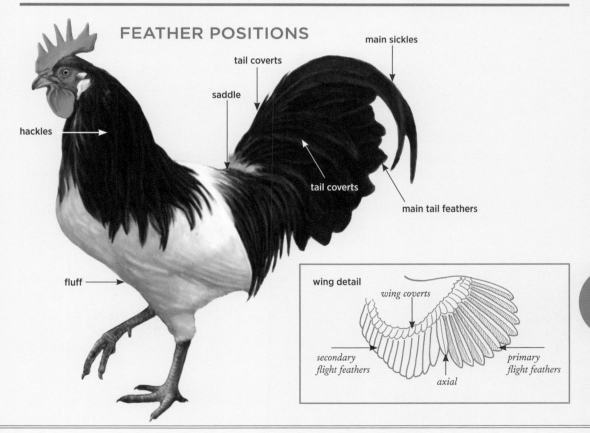

hackles

saddle

tail coverts

main sickles

tail coverts

main tail feathers

fluff

wing detail

wing coverts

secondary flight feathers

axial

primary flight feathers

F

feathering \ The sum total of a chicken's plumage. \ The feathers covering a particular part of a chicken, such as its leg feathering.

Feather Leg (FL) class \ One of the groupings into which the American Poultry Association and American Bantam Association organize bantam breeds. This class includes the feather-legged breeds: Belgian Bearded d'Uccle, Booted, Brahma, Cochin, Faverolle, Frizzle, Langshan, Silkie, and Sultan.

feather-legged \ Having feathers growing down the outsides of the shanks and on the outer toe or on both the outer and middle toes.
[Also called: shank feathered]

featherless chicken \ A hybrid developed in Israel by crossing a Naked Neck with a normal broiler. These chickens have only a few wisps of feathers on their pink skin, allowing them to waste little energy growing feathers instead of meat. Compared to regular broilers, featherless chickens grow faster in hot climates, where growth rate is otherwise limited because nutritional energy must be restricted to prevent broilers

from overheating under full plumage. Featherless chickens require shade to prevent sunburn and in a cold region would need heated housing.

feather loss \ *See: brood patch; feather picking; molt; treading*

feather mite \ *See: northern fowl mite*

feather out \ The process whereby a chick grows its first full set of feathers, usually at about six weeks of age but sometimes later depending on breed.

feather picking \ A form of cannibalism most common when chicks are feathering out or mature birds are molting and the newly emerging feathers attract the attention of flock mates. Emerging feather quills are filled with blood, and once chickens get a taste, they want more. Feather picking has many causes, often working in combination. Conditions that can trigger it include:

- Crowding, especially in fast-growing chicks that fill the available space and can't get away from each other
- Bright lights left on 24 hours a day
- Too-warm housing temperature
- Inadequate ventilation
- Too few feeders and drinkers
- Feed and water stations left too close together as chicks grow
- Diet too low in protein
- Insufficient opportunities to engage in normal chicken behavior, especially scratching and pecking the ground

feather sexing \ Separating newly hatched hybrid chicks by gender, based on the growth rate of their wing

featherless chicken

F

feathers. Crossing a slow-feathering hen (such as a Rhode Island Red) with a rapid-feathering cock (such as a white Leghorn) yields slow-feathering cockerels and rapid-feathering pullets. The chicks are all the same color but may be sexed by pullets' well-developed wing feathers at the time of hatch.

Feather sexing is commonly used in the broiler industry — where white-feathered birds are preferred — so the slow-growing pullets may be raised apart from their faster-growing brothers. The difference in the growth of the primary feathers is seen only between one and three days of age, after which the cockerels' wing feathers catch up with the pullets' and they all look alike.

feather tract \ An area of skin on which feathers grow. Unlike a mammal's fur, feathers do not cover the chicken's entire body surface but grow in narrow, symmetrical tracts. A chicken has 10 distinct feather tracts — head, neck, shoulder, wings, breast, back, abdomen, rump, thigh, and legs — from which the

feathers fan out to cover the chicken's body. The feathers' follicles are linked by a network of tiny muscles that allow a chicken to raise and lower its feathers; for instance, to trap warm air by puffing out the feathers in cold weather.

The feather tracts (technically, pterylae) are separated by featherless areas (apteria) that may contain some down. These bare areas facilitate cooling when a chicken holds out its wings in hot weather; otherwise, no one is exactly sure why chickens and other birds grow feathers in specific tracts.

fecal \ Pertaining to feces.

fecal test \ Examination of fresh chicken droppings, typically by a veterinarian, to diagnose the presence or absence of a medical condition. A fecal test is most commonly used to detect the need for deworming but may also be used to identify the cause of a digestive disorder and thus determine the appropriate treatment.

feather sexing

pullet

cockerel

feather tracts

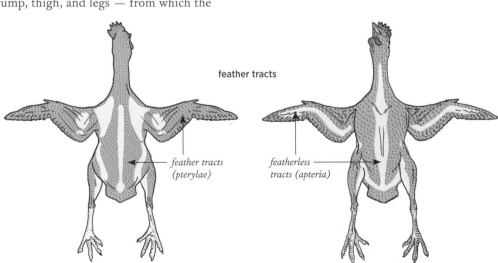

feather tracts (pterylae)

featherless tracts (apteria)

F

feces \ *See: droppings*

fecund \ Capable of laying a large quantity of eggs, a trait that varies from breed to breed and from strain to strain within a given breed. The most fecund hens have been selectively bred for their laying ability. *See also: rate of lay*

feed \ Appropriate food for chickens, the sum total of which makes up their daily ration. Chicken feed might be commercially prepared, homegrown, or a combination of both, provided it is nutritionally balanced to meet the chickens' needs based on the season, the temperature, and the bird's age, size, weight, and rate of lay.

feed amount \ The amount a chicken eats. It varies with the season and temperature, as well as with the bird's age, size, weight, and rate of lay. Since chickens eat to meet their energy needs, the amount of a particular ration a bird will eat depends also on the ration's energy density. A chicken fed the same ration year-round will eat more during cold weather, because it needs more energy to keep its body warm. As some extremely general guidelines, expect to feed:

- Each hybrid broiler about 10 pounds (4.5 kg) total feed to reach butchering age
- Each heritage broiler about 20 pounds (9 kg) total feed to reach butchering age
- Each light-breed pullet about 15 pounds (6.75 kg) total feed to reach age of lay
- Each heavy-breed pullet about 30 pounds (13.5 kg) total feed to reach age of lay
- Each mature bantam about ½ pound (0.25 kg) of feed per week
- Each mature light-breed chicken about 2 pounds (1 kg) of feed per week
- Each mature midweight dual-purpose chicken about 3 pounds (1.5 kg) of feed per week
- Each mature heavy-breed chicken about 4 pounds (2 kg) of feed per week

feed conversion \ The pounds (kilograms) of feed needed to produce a dozen eggs or increase a broiler's weight by 1 pound (0.45 kg). A good conversion rate for layers is 4 pounds (1.8 kg) of feed per dozen eggs produced. A good conversion rate for commercial hybrid broilers is 2 pounds (0.9 kg) of feed per pound (0.45 kg) of weight gained, which amounts to about 3 pounds (1.4 kg) of feed per pound of dressed weight. A typical conversion rate for pastured broilers is 4 pounds (1.8 kg) of feed per pound (0.45 kg) of weight gained, which amounts to about 6 pounds (2.7 kg) of feed per pound of dressed weight.

feed

crumbles   *pellets*

F

feed cost \ The item that accounts for most of the expense of maintaining a layer flock. Cost may be minimized by guarding against feed wastage and by seeking economical feed sources, including natural forage.

feeder \ The appliance in which rations are placed for chickens to eat. Feeders come in many different styles, the two most common being a hanging tube and a long trough. If you feed free choice, put out enough feeders so at least one-third of your chickens can eat at the same time. If you feed on a restricted basis, put out enough feeders so the whole flock can eat at once.

Allow each mature chicken at least 1.5 inches (3.75 cm) of space around a tube feeder or 1 inch (2.5 cm) of space along a trough feeder. If the chickens can eat from both sides of the trough, count both sides in your calculation.

feeder, chick \ Chicks look for things to peck soon after they hatch. To help them find feed, sprinkle a little starter ration on a paper towel, paper plate, or sheet of aluminum foil. Place the starter no farther than 24 inches (60 cm) from the heat source but not directly under it. As soon as the chicks are all eating, remove the feed-covered paper and replace it with a shallow lid or tray, such as a shoe-box lid. When the chicks start scratching feed out of the tray, switch to a regular chick feeder, available in several styles from a farm store or poultry supplier.

A JAR FEEDER consists of a feed-filled glass or plastic jar that screws onto a round base with holes through which the chicks eat. The smallest size holds a quart (1 L), and its small footprint makes it ideal for use in a brooder with limited space. A similar style of round feeder consists solely of an enclosed base, with no provision for screwing onto a jar. It has the same footprint but holds less feed. Chicks quickly outgrow both styles. Allow one hole per chick.

A TROUGH FEEDER is made of either plastic or metal and comes in various lengths from about a foot (30 cm) on up. If the trough has a lid with holes through which the chicks eat, allow one hole per chick until the chicks outgrow the feeder. If the trough has no lid, allow 1 linear inch (2.5 cm) per chick up to 3 weeks of age, 2 linear inches (5 cm) to 6 weeks of age, and 3 linear inches (7.5 cm) to 12 weeks. Count both sides if chicks can eat from either side.

A TUBE FEEDER is made of plastic and comes in several sizes, the smallest holding just under 2 quarts (almost 2 L). It is designed to hang from a rope or chain, to be raised as the chicks grow. Keeping the base about the height of the chicks' backs discourages beaking out and helps minimize wasted feed. To determine how many chicks aged six weeks or under may be fed from one tube feeder, measure the feeder's base diameter in inches (or centimeters), multiply by 3.14, and divide by 2 (in metric divide by 5).

tube feeder

chick feeder

F

two weeks after it was mixed; nutritional deficiencies result from feed stored so long the fat-soluble vitamins are destroyed by oxidation. If you prefer not to feed a commercial ration, supplement your flock's diet with animal protein and a vitamin/mineral supplement, starting six weeks before hatching eggs will be collected. If the flock has been getting scratch grain, ensure the proper proportion of protein to carbohydrates by reducing the grain ration about a month before the hatching season begins.

feeding broilers \ Commercial rations may not contain sufficient nutrients to sustain the rapid growth rate of Cornish-cross broilers, and as a result they develop leg problems. Lameness may be minimized by supplementing starter or grower ration with a vitamin/mineral mix, as well as by reducing the growth rate using one or more of the following measures:

- Reduce the number of hours the broilers are under lights, thus reducing the amount of time they spend eating
- Altogether withhold feed overnight
- Feed a nonpelleted ration to increase the amount of time required to eat the same amount of feed
- Substitute up to 30 percent of the total ration with whole wheat to reduce the rate of digestion

Lameness problems may be avoided by raising a heritage breed or hybrid that does not achieve the exaggerated growth rate of industrial-production broilers. Another option is to raise broilers on pasture. Some growers of pastured or range-fed broilers feed a high-protein

feeding behavior

feeding behavior \ Normally, scratching in the soil for food items to peck and swallow. If an item is too big to swallow whole, the chicken might break it up through head shaking or beak beating or might engage other chickens in food running or tidbitting. Feeding behavior also includes cleaning particles of food from the head and beak, which a chicken does by scratching its head and beak with a claw or by wiping its beak against the ground.   *See also: beak beating; food running; tidbitting*

feeding breeders \ The same ration that promotes good egg production won't necessarily provide embryos and newly hatched chicks with all the elements they need to thrive. To improve hatching success, feed a higher-protein layer ration, a breeder ration, or a game-bird ration starting two to four weeks before eggs are collected for hatching. Be sure the ration is fresh, and use it within

F

diet for fast, economical growth. Others favor a high-energy diet that results in more flavorful meat.

Muscle (meat) growth relies on protein. Pastured broilers may be fed a ration of up to 30 percent protein. Using a ration with as little as half that amount of protein can reduce the growth rate by as much as 50 percent. Broilers taking longer to grow eat more total ration, so feeding the less-expensive low-protein ration can cost more than feeding a pricier high-protein ration. *See also: Label Rouge*

feeding chicks \ Newly hatched
chicks can survive for two to three days without drinking or eating, as their bodies are still absorbing yolk reserves. However, the sooner they eat, the less stress they will experience and the better they will thrive. To minimize pasting, chicks should always drink before they start eating. Mashed hard-boiled egg combined with oatmeal makes an excellent and time-honored starter.

Some farm stores offer a starter ration, a separate grower or finisher ration for broilers, and a grower or developer ration for layers and breeders. Switch from one ration to another as indicated on the labels, making the change gradually by combining the old ration with increasing amounts of the new ration to avoid digestive upset. Where only one all-purpose starter or starter-grower ration is available, continue using it until broilers reach butchering age and layers/breeders are ready for the switch to lay ration.

Never feed lay ration to chicks, as its high calcium content can interfere with bone formation and cause weak legs, kidney damage, and possibly death.

feeding cocks \ In most backyard
situations roosters eat the same ration put out for hens. However, outside of breeding season, cocks require only a 9 percent protein maintenance ration. Mature cocks housed separately from hens may be fed more economically by using Pearson's square to reformulate their ration to reduce the protein content.

feeding hens \ Chickens eat to
meet their energy needs, so they eat less in summer than in winter, when they need extra energy to stay warm. If the summer ration contains the same amount of protein as the winter ration, hens will get less total protein in summer and won't lay as well. In warm climates, in addition to regular 16 percent lay ration, some farm stores offer a ration containing 18 or 20 percent protein for use when high temperatures cause hens to eat too little to sustain laying.

When the weather is cold and days are short, hens may not get enough to eat to both maintain body heat and continue laying well. Lights in the housing may be used to increase daylight hours and give them more time to eat. Additional energy may be furnished by supplementing the regular ration with milk and/or scratch grains. Feeding hens table scraps and surplus milk products year-round may reduce the cost of egg production but may also reduce egg production itself, unless care is taken to maintain dietary balance.

F

feeding pullets \ Pullets that are not in good flesh when they start laying cannot obtain enough dietary protein to both continue growing to maturity and lay eggs. As a result they lose weight and produce small eggs, as well as fewer of them. Pullets that grow too rapidly and are fat when they start laying also tend to lay fewer, smaller eggs and may have trouble laying them.

Start pullets as chicks with enough protein to give them a good beginning, then back off a bit to give them time to mature without getting fat before they begin to lay. A good way to reduce protein is to add oats gradually to the grower or developer ration. When pullets are eight weeks of age, use Pearson's square to reformulate the ration to reduce protein steadily to a level of between 14 and 16 percent.

As the pullets reach laying age, their need for dietary protein goes up. Leghorn-type breeds will start laying at about 20 weeks, other breeds at 22 to 24 weeks. About 2 weeks before the age of lay, gradually reduce the oats to a protein level of between 16 and 18 percent, then steadily switch over to a layer ration to provide the higher calcium and other nutrients hens need for laying. Do not feed layer ration to pullets before they reach the age of lay, as the high calcium content can cause serious kidney damage.

feeding setters \ A setting hen eats about one-fifth the amount she normally eats and on some days doesn't eat at all. When she does get off the nest to eat, she'll take in more water than solid food. During the 21 days of incubation, she'll lose as much as 20 percent of her body weight. A setting hen therefore must start out at the peak of health with a good layer of body fat, as indicated by a creamy or yellow hue to the skin; a hen with reddish- or bluish-looking skin has insufficient fat reserves to see her through until the chicks hatch.

While the hen is setting, feed scratch grain instead of layer ration to help her maintain weight and also to keep her droppings solid — as well as easier to clean out of the nest and less likely to stick to the eggs — should she accidentally poop in the nest. Put grain and water near the nest, but don't worry if the hen doesn't eat for the first few days. After that she should get off the nest for 15 or 20 minutes at about the same time each day to drop a broody poop, take a deep drink, grab a few kernels of grain, maybe take a quick dust bath, before returning to the nest.

feeding show chickens \ Experienced exhibitors feed their show birds a custom diet based on, among other things, the desired weight for their breed and the effects of certain feeds on plumage color. A good starting place is basic breeder ration.

Some exhibitors add corn to reduce egg production and improve feather quality, but avoid feeding yellow corn to white varieties or those with white earlobes, since corn tends to run to fat, and its pigment gives white birds a brassy hue. Whole oats, fed free choice in a separate hopper from the breeder ration, will improve feather quality without making a bird fat.

To stimulate natural oil production and promote bloom, include a small

F

amount of oil-rich feed, such as safflower seed, sunflower seed, flaxseed, or linseed meal. To further enhance the gloss of a variety with black or red plumage, feed a tiny bit of quality canned cat food daily.

feed mill \ A grinder, such as a heavy-duty flour mill, that produces a coarse grind to improve the digestibility of corn, beans, and large grains. Mills come in inexpensive hand-crank versions or less taxing motor-driven versions.

feed storage \ From the time a ration is milled and mixed, it begins losing nutritional value through oxidation and other aging processes. Use any prepared feed within about four weeks of being milled. Allowing a week or two for transport and storage at the farm store, buy only as much as you can use within two to three weeks.

Store bagged feed off the floor on pallets or scrap lumber, away from moisture. After opening a bag, pour the feed into a clean plastic trash container with a tight-fitting lid, and keep it in a cool, dry place, out of the sun. A plastic container works better than a metal can, which sweats in warm weather, causing feed to get moldy. A closed container slows the rate at which feed goes stale and keeps out rodents.

Use up all the feed in the container before opening another bag; never pour fresh feed on top of old feed. If you have a little feed left from a previous batch, pour it into the container's lid, pour the fresh bag into the container, and put the older feed on top so it will be used first.

feed mill

fence/fencing \ A sturdy fence keeps chickens in and predators out. The ideal chicken fence is made from tightly strung, small-mesh woven wire. Chain link is expensive but virtually maintenance free and keeps most predators out. Small-mesh welded or woven wire also works well. Chicken wire works in the short term, but it deteriorates too rapidly to be practical.

Electrified scare wires prevent predators from climbing over the fence, an apron thwarts diggers, and wire across the top of a small run keeps out flying predators. In populated areas where an agricultural-style fence or electrified wires are not permitted, a solid wooden fence, a high picket fence, or an attractive rail fence backed with a less visible wire mesh are some alternatives.

The fence must be at least 4 feet (1.2 m) high, higher if you keep a lightweight breed that likes to fly. Bantams and young chickens of all breeds are fond of flying.

F

fencer \ Electric fence controller.

fence skirt \ *See: apron*

fertile \ Describes an egg that has been inseminated, making it capable of producing a chick if incubated under appropriate conditions. An egg must be fertile in order to hatch, but not all fertile eggs do hatch. A fertile egg is fine for eating and has the same nutritional value as an infertile egg, since sperm contribute an insignificant amount of nutrients. \ A laying hen that has been exposed to a cock and therefore produces fertile eggs. Whether or not hens are actually inseminated depends on the cock's fertility and the mating ratio. \ A virile cock that inseminates hens or from which viable sperm is collected for artificial insemination, for the production of hatching eggs. The production of semen may be stimulated to improve fertility by exposing cocks to 14 hours of light, artificial and daylight combined, for four to six weeks prior to collecting eggs for hatching.

fertility \ When a hen is inseminated, sperm travel quickly up the oviduct to fertilize a developing yolk. If the hen laid an egg shortly before, the mating will likely fertilize her next egg. The number of additional eggs that will be fertilized by the mating varies with the hen's productivity and breed. The average duration of fertility is about 10 days. Highly productive hens remain fertile longer than hens that lay at a slower rate, and single-comb breeds remain fertile longer than rose-comb breeds — possibly as long as a month, but that's pushing it.

Eggs laid after a hen has been mated with a new cock are more likely (but not guaranteed) to be fertilized by the new male than by the old one. Allow at least two weeks, however, to be reasonably certain eggs collected for hatching have been fertilized by the new mating.

fertility, weak \ A term typically used to describe embryo death occurring before incubation began, which in fact may have had nothing to do with fertility. Early embryo death may occur due to deficiencies within the egg, such as might be caused by disease or dietary issues; because the egg was incubated for a brief time (for instance, by a hen laying subsequent eggs) and then left too long before incubation resumed; or the egg was subjected to extreme cold or warm temperatures before being placed in an incubator or under a setting hen.

fertility of show birds \ Frequently showing breeders can result in poor fertility and inferior chicks. Birds become stressed by travel, inconsiderate spectators, peculiar feed and feeding schedules, and perhaps lack of water due to oversight or because the birds don't like the unusual taste. Lack of water is particularly a problem for layers. Hens, in general, are more greatly stressed by showing than are cocks, and older birds of either sex are more strongly affected than younger ones. Keep valuable breeders away from the showroom and the consequent exposure to stress and potential disease. Alternatively, minimize the risk by hatching a hefty number of chicks before showing breeder stock.

CAUSES OF LOW FERTILITY

Aside from problems related to inbreeding depression, management factors (rather than inheritance) are more likely to be the cause of low fertility among hatching eggs. The many possible reasons for low fertility include the following:

- The flock is too closely confined
- The weather is too warm
- Breeders (both cocks and hens) get fewer than 14 daylight hours
- The cock has an injured foot or leg
- Breeders are infested with internal or external parasites
- Breeders are diseased; especially troublesome to fertility are chronic respiratory disease, infectious coryza, infectious bronchitis, Marek's disease, and endemic (mild) Newcastle disease
- Breeders are too young or too old
- Breeders are stressed due to excessive showing
- Breeders get too little protein or are otherwise undernourished
- The mating ratio is too high or too low

Breed-related mechanical problems may also result in low fertility. Such mechanical problems include these:

- **COMB SIZE.** Breeds with large single combs have trouble negotiating feeders with narrow openings, and the resulting nutritional deficiency affects fertility.

- **CRESTS.** Houdans, Polish, and other heavily crested cocks may not see well enough to catch hens. A quick fix is to clip back their crest feathers.
- **HEAVY FEATHERING.** Brahmas, Cochins, Wyandottes, and other heavily feathered breeds have trouble mating. Fertility may be improved by clipping vent feathers.
- **FOOT FEATHERING.** Booted bantam cocks and males of other breeds with heavy foot feathering have trouble getting a foothold when treading hens.
- **RUMPLESSNESS.** Araucanas (and occasionally birds of other breeds) have no tail to pull the feathers away from their vents during mating. The quick fix is to clip the vent feathers of both cocks and hens, with more attention to the feathers above the hens' vents and those below the cocks' vents. A better solution is to select breeders with the least vent feathering.
- **HEAVY MUSCLING.** Cornish cocks and other heavy-breasted males have trouble mounting hens because of the wide distance between their legs. These breeds are typically bred through artificial insemination.

Fertility issues may be genetically related. Cocks that sport rose combs generally have lower fertility than single-comb cocks. Any breed, or strain within a breed, that is highly inbred because of a low population tends to be low in fertility.

F

finish \ The condition of a mature chicken in full feather, displaying the proper color and ideal weight for its breed and variety, being clean and in perfect health, and having the radiant glow of bloom. \ The amount of fat beneath a broiler's skin as an indication of flavor. To assess finish, spread the breast feathers and examine the skin. A creamy or yellow color indicates good finish; a reddish or bluish color indicates too little fat.

fixed housing \ A shelter constructed in a permanent location, as compared to portable housing periodically moved around a garden or pastureland. Because the yard outside a fixed shelter readily turns to packed dirt or mud and the accumulated manure eventually leads to disease, permanent housing works best for a small flock, in a dry climate, and with a good system of yard rotation.

flat \ A stackable egg tray made of plastic or paper pulp, usually designed to hold 30 eggs. \ A wingette, the outer two sections of the wing, prepared and served as finger food. It's called a flat because, when separated from the rest of the wing at the elbow, a wingette lies flat.

flat shins \ Undesirable shanks that are flat in front, rather than being properly rounded.

flatworms \ Parasitic tapeworms (cestodes) and flukes (trematodes). Flatworms are far less common in chickens than are roundworms. *See also: worm*

flaxseed \ The seed of *Linum usitatissimum*, fed to hens as an alternative to pasture to increase the omega-3 fatty acids in their eggs. Besides being high in omega-3s, flaxseed is high in protein and contains a large number of vitamins and minerals. But feeding flax as more than 10 percent of the total diet can result in fishy-tasting eggs.

Hens don't normally care much for flaxseed. Crushing the seeds before adding them to the ration makes them more likely to be eaten. Alternatively, use food-grade linseed oil. Either way, to avoid rancidity, mix in only as much into their ration as your hens will eat within a day or so.
[Also called: linseed]

flats

fleshing \ Mature growth of the breast, legs, and thighs.

flies \ Insects of the order Diptera, characterized by having one pair of wings for flight and a second small pair for balance. Flies that bother chickens fall into two categories:

BITING FLIES have sucking, and often also piercing, mouthparts. The main ones that bother chickens are blackflies (also known as buffalo gnats or turkey gnats) and biting gnats (also known as midges, no-see-ums, punkies, and sand flies). Bites from these flies cause irritation and can spread diseases. Control of biting flies is difficult and primarily involves keeping chickens away from streams and stagnant water.

FILTH FLIES, including the common housefly, don't bite, but they do transmit tapeworms (to chickens that eat infected flies) and spread diseases on their feet. Filth flies breed in damp litter and manure. Their control involves keeping litter dry by fixing leaks and seepage and improving ventilation. Properly managed bedding and manure allowed to accumulate during fly season develop a natural population of fly predators. Chickens also help control flies by eating mature flies and their larvae.

If flies get out of control, the use of insecticides can lead to a resistant fly population. Instead, set out fly traps or good-quality flypaper, or introduce natural fly predators. *See also: fly parasites/ predators*

flight coverts \ The rows of contour feathers covering the bases of the wings' flight feathers. The *Standard* definition limits flight coverts to only the primary coverts.

flight feathers/flights \ The large, stiff feathers of the wing — consisting of the primary flight feathers and the secondary flight feathers — which are aerodynamic and therefore help a chicken fly. Although technically the flight feathers also include a chicken's main tail feathers, in the world of chickens the flight feathers are considered to be wing feathers, and the *Standard* definition further narrows them down to only the wing's primaries.

flights

flighty \ Easily excited.

flip-over disease \ *See: sudden death syndrome*

flipper \ A victim of sudden death syndrome, also known as flip-over disease.   *See also: sudden death syndrome*

flock \ A group of chickens living together.

flock breeding \ Collecting hatching eggs from a group of chickens, which may include more than one breeder cock, without keeping track of which eggs come from which specific matings. Flock breeding emphasizes a flock's overall production performance, as opposed to emphasizing characteristics of superior individuals.

F

flock history \ A record that includes all information pertaining to a group of chickens, including their age(s), familial strains, medications and vaccinations, contact with other flocks (by means of moved birds or human visitors), past disease issues, and the circumstances surrounding any death(s). This history may be used to trace back the source of problems (such as those involving health or breeding) that may arise in the future.

fluff

flogging a hen \ *See: treading*

floor egg \ An egg laid on the floor rather than in a nest. Floor eggs, often deposited by pullets just starting to lay, tend to get dirty or cracked, making them unsafe to eat. Pullets may lay on the floor because:

- Nests were not provided in time for the pullets to get used to them before they started laying
- Nests are too high for the pullets to reach
- Too few nests are furnished for the number of pullets (provide at least one nest for every four layers)
- The nesting area is too bright, causing pullets to seek out darker corners for laying

flooring \ The main considerations in deciding what type of flooring to use in a chicken shelter are cost, ease of cleaning, and resistance to rodents and predators. Any flooring needs to be covered with bedding.

- **DIRT** is cheapest but cannot be made rodentproof. In warm weather it helps keep birds cool, although in cold weather it draws away heat.
- **WOOD** is popular but hard to clean, and it attracts rodents, which take up residence underneath, unless the floor is at least 1 foot (0.3 m) off the ground.
- **CONCRETE,** well finished, is expensive, but it's easy to clean and keeps out rodents and predators.

floor space \ *See: space requirements*

fluff \ The portion of a feather at the lower end of the shaft, which has no hooks to form a smooth web and is therefore soft and downy. This part of the feather is usually not visible unless the feathers are deliberately disturbed. \ The soft, downy feathering on the inner side of a chicken's lower thighs and on the abdomen. \ An accumulation of down sheaths from newly hatched chicks.

fluke \ A broad, leaf-shaped parasitic trematode flatworm that attaches itself either inside a chicken's body or beneath its skin. Flukes are a problem primarily in swampy areas and where sanitation is poor. *See also: worm*

fly parasites/predators \ Tiny wasps that live on or near manure and other decaying matter. They are not really parasites, even though they live off flies, and are not really predators, even though they prey on flies, but technically are parasitoids, as they live parasitically off flies and eventually kill them.

These wasps do not bite or sting humans, chickens, or other animals, and neither do they attack mature flies.

F

Instead, a female wasp seeks out fly pupae and deposits from one to a dozen eggs (depending on the wasp species) into each pupa. Moving from pupa to pupa, she lives just long enough to deposit all of her 50 to 100 eggs, which develop into wasp larvae that feed inside the host fly pupa, thereby killing it. In 14 to 28 days mature wasps emerge from the fly pupa, and the females begin searching out new host pupae to lay eggs in.

Under favorable conditions, the wasps populate on their own. To assess their effectiveness, collect fly pupae from different spots, put them in a jar, and wait to see whether flies or wasps emerge. Since flies reproduce faster than parasitoids, sometimes they become too numerous for the wasps to control. In such an event, recruit wasps may be purchased from biological pest control suppliers and released to augment the local population.

fomites \ Objects such as cages, feed sacks, clothing, shoe soles, and vehicle tires that may carry and transmit infectious organisms. From the Latin word *fomes*, meaning tinder — as in, the infectious organism so transmitted can spark a disease outbreak.

food call \ A high-pitched sound repeated rapidly by a hen to bring her chicks' attention to something tasty to eat; also used by a cock to call hens when he finds some tasty tidbit.

food running \ A habit of chickens (that's so instinctive even baby chicks do it) of grabbing a large food item in the beak and running with it. One theory is that the running chicken wants to keep the food for itself and is seeking a place away from the others where it can break apart the food in private. Another theory is that the running chicken wants to attract the attention of other chickens, which give chase and engage in a tug-of-war over the item, for the purpose of breaking up the food into pieces small enough to swallow.

foot \ The chicken's foot is made up of 16 bones that provide the flexibility needed to withstand the stresses of running, balancing on a roost, taking off in flight, and landing (although heavy breeds are not well adapted to jumping down from heights). The bones of the ankle and foot are fused and elongated to form the shank, technically making the shank part of the foot rather than part of the leg.   *See also: toe*

shank

hind toe

toes

claw

F

foot pad dermatitis \ An abscess in the pad on the bottom of a chicken's foot, caused by the bacterium *Staphylococcus aureus*. The infection most commonly affects maturing heavy-breed cocks in one foot or both and is usually discovered while one is trying to determine why a cock walks with a limp and sits on his hocks while at rest. The condition may be prevented by keeping heavy breeds off wire or hard-packed flooring to avoid irritating the bottom of the foot and by removing high perches from which the birds may jump down and injure a foot, opening the way to bacterial infection.
[*Also called: bumblefoot*]

foragers \ *See: self-sufficient breeds*

forced-air incubator \ *See: circulated-air incubator; incubator*

forced molt \ An artificially stimulated molt, designed to force a population of laying hens to molt together and resume egg production more rapidly than they would if they molted naturally. Forced molting involves inducing severe nutritional and environmental stress and if not done according to a precise plan can result in dead hens. It is therefore generally considered cruel and inhumane.
[*Also called: controlled molt (an industry euphemism)*]

forked worm \ *See: gapeworm*

fossil flour \ *See: diatomaceous earth*

foster hen \ A hen that raises hatchlings she didn't incubate herself, which may be poultry other than chicks. A hen is most likely to accept foster hatchlings if she's been seriously setting for a couple of weeks but not necessarily the entire 21 days. Similarly, the hatchlings must be not much more than a day old and still receptive to accepting (or imprinting) the hen as Mom. To increase the chances that both parties will accept each other, slip the hatchlings under the hen at night. Be prepared, in the event things don't work out, to gather up the babies and brood them yourself.

fowl \ Poultry. \ A stewing hen. *See also: poultry*

fowl catcher \ *See: catching hook*

fowl pox \ A viral disease, unrelated to chicken pox in humans, that causes scabby skin, fever, and loss of appetite. It is spread by blood-sucking insects and through injuries, such as those resulting from peck-order fighting. A vaccine is available but should be used only if a flock, or the area where the flock lives, has a pox problem; then all chickens in the flock must be vaccinated and receive yearly boosters.

frazzle \ *See: curly*

free choice \ Available to chickens at all times, so individuals can take as much as they need. Drinking water should always be offered free choice. Supplements should be offered free choice to allow chickens to vary their intake according to the amount their

F

bodies need at various stages in their lives. Feed is usually offered free choice, to ensure no chicken goes hungry.

free range \ Technically, means chickens are not confined, but in practical terms means chickens are allowed to roam at will within a fenced outdoor area. Federal regulations say free-range chickens must be allowed access to the outdoors but don't indicate they actually have to go outside. Most people consider "free range" to mean chickens have continuous access to fresh air, sunshine, and exercise as opposed to being confined in cages. *See also: pastured/pasture raised; range feeding*

free roaming \ Not confined to a cage. The term usually indicates the birds are free to move around within the confines of a building but sometimes is used synonymously with free range.

Frizzle \ A chicken with frizzled feathers. At one time Frizzles were considered a distinct breed that had a single comb, could be either large or bantam, and came in both clean-legged and feather-legged varieties, as well as in a limited number of solid colors. Today a Frizzle is considered to be any breed or variety with frizzled feathers and is named after the breed that has been frizzled — Cochin Frizzle, Polish Frizzle, and so forth.

frizzled \ A genetic condition of having feathers that lack a smooth web and that curl outward and backward. A frizzled chicken has a unique permed look, has less protection from the elements than a chicken with smooth feathers, and is flightless.

Frizzledness is genetically dominant and may be introduced into any breed by mating a frizzled chicken with a smooth feathered chicken. Half the offspring will be nonfrizzled and half frizzled, although the frizzledness won't be apparent until the chicks feather out. Breeding two frizzled chickens together is frowned on — half the chicks will be frizzled, 25 percent will be smooth feathered, and 25 percent will be undesirable curlies. *See: curly*

frizzled

frostbite \ Injury to body tissues, most commonly of the comb or wattles but also possibly of the toes, caused by exposure to freezing temperatures and sometimes resulting in gangrene. Large-comb breeds are more likely to suffer frostbite than breeds with smaller combs. Cocks are more likely than hens to suffer frostbite, since they have larger combs and, unlike hens, usually don't sleep with their heads tucked under a wing. To minimize the possibility of frostbite:

- Reduce humidity by removing damp litter and improving ventilation
- Install perches in the least drafty part of the housing
- Coat combs and wattles with petroleum jelly as insulation against frozen moisture in the air
- Furnish a source of gentle heat over roosts that are more than 2 feet from the ceiling

frosting \ An undesirable pale or faded line around the edge of a black laced or spangled feather.

fryer \ *See: broiler*

F

gait scoring \ A method of evaluating the ability of a chicken to walk, originally developed for breeder flocks but now used primarily to assess the welfare of commercially raised broilers. The six-point Kestin score developed in England has been simplified into a three-point U.S. gait-scoring system.

gamy tail

G

Galliformes \ An order of land-dwelling birds that nest on the ground and raise large broods. They typically have large bodies with short wings, and their legs and feet are designed for running and scratching up food. This order includes chickens, pheasants, turkeys, guinea fowl, and other domestic poultry and game birds.

gallinaceous \ Resembling poultry; relating to birds of the order Galliformes. From the Latin word *gallina*, meaning hen.

Gallus domesticus \ A bird of the genus *Gallus* (encompassing all chickens) and the species *domesticus* (domestic chickens). From the Latin word *gallus*, meaning cock.

game breeds \ Chickens originally developed for the purpose of cockfighting. Game breeds include American Game bantam, Aseel, Malay, Modern Game, Old English Game, and Shamo. Although the Modern Game was deliberately created for exhibition only, it was selectively bred from the Old English Game.

game class \ A subgroup of the All Other Standard Breeds (AOSB) class, one of the six groupings into which the American Poultry Association organizes large chicken breeds and also one of their classifications for bantam breeds. The game class includes Modern Game and Old English Game. The American Bantam Association additionally recognizes American Game bantams, grouping them with Old English Game and giving Modern Game a class of its own.

gamy tail \ A tail that is held tightly folded and tapering toward the end. Gamy tail is characteristic of Modern Games and Cornish and to a lesser degree of Malays but is undesirable in other breeds.

gapes \ Disease resulting from being infected with gapeworm.

gapeworm \ A red, fork-shaped roundworm (nematode) that invades the windpipe, causing an infected chicken to gasp (gape), cough, and shake its head in an attempt to dislodge it. These parasites are quite serious in young birds, since they can cause death through strangulation, although mature chickens can become resistant. It is called a forked worm because each female has a male gapeworm permanently attached, forming the letter Y; it is sometimes called a red worm because the female is blood red.

gardening with chickens \

The synergistic practice of using chicken manure to fertilize a garden, feeding garden gleanings to the chickens, and letting the chickens scratch in the soil to reduce insect pests and weed seeds during fallow times. A well-managed system can work to the benefit of both the chickens and the garden.

gene \ A unit of heredity on a chromosome that transfers characteristics from a parent to offspring. Each chick acquires two sets of genes, one from its sire and one from its dam. The genes match up into pairs of like function: for example, controlling comb style or feather color.

Paired genes that are identical to each other are homozygous; paired genes that differ from each other (alleles) are heterozygous. Each gene can be either dominant or recessive. If a dominant pairs with a recessive, the dominant trait overshadows or modifies the recessive trait, and the dominant trait prevails.

A homozygous chicken is more likely than a heterozygous chicken to have a large number of matched pairs, so the more inbred a bird is, the more likely it is to display recessive traits. In a population of heterozygous chickens, recessives can remain hidden by dominants, to pop up in the future when two genes pair that control the same recessive trait.

Some traits are controlled not by dominant or recessive genes but by combinations of genes. An example is rumplessness, a genetically complex feature of Araucanas determined by the interaction among many different genes.

Some genes are lethal. A chick that acquires the same lethal gene from both parents dies early, often in the embryo stage.

gene pool \ The complete collection of all alleles in the genetic makeup of every chicken in a given population. A large gene pool has extensive genetic diversity; a small gene pool has low genetic diversity that can lead to genetic erosion.

general scale of points \ A
list of values assigned by the *Standard of Perfection* to a chicken's every feature — size, condition and vigor, comb, crest, beak, skull and face, eyes, wattles, beard, earlobes, neck, back, tail, wings, breast, body and fluff, legs and toes, symmetry,

and carriage or station. A specific number of points is allowed for each trait, with deductions (or cuts) made for such things as incorrect weight, missing tail feathers, off-color eyes, and other defects. In theory, a licensed judge is supposed to appraise show birds according to the scale of points, but in today's fast-paced world, judging by points and cuts takes too long.

genetic \ Pertaining to genes.

genetic diversity \ Degree of variation at the level of individual genes. The greater the genetic diversity among chickens, the more robust is the population and the better able it is to resist disease and other stresses that change as the environment changes. Low genetic diversity results in reduced viability and can lead to genetic erosion.

Because poultry sperm and embryos cannot be preserved easily through freezing, the primary way to maintain genetic diversity is through living flocks. In a given breeding population, a greater degree of genetic diversity is retained by keeping as future breeders the best individuals from each mating; reduced genetic diversity results from retaining the best representatives of the overall population.

genetic erosion \ The diminishing of an already limited gene pool, resulting in an escalation in the loss of genetic diversity. Genetic erosion occurs when poultry keepers lose interest in keeping certain breeds (which may become extinct) or in retaining existing characteristics within a breed (which are then lost to future generations of the breed).

German Spitzhauben \ *See: Spitzhauben*

germinal disc/disk \ *See: blastodisc/blastodisk*

giblets \ Edible organs consisting of the liver, heart, and gizzard.

gizzard \ A chicken's mechanical stomach, which makes up for the bird's lack of teeth by pulverizing plant fibers and whole grains for digestion. Lying between the true stomach and the small intestine, the gizzard consists of strong muscles surrounding a tough pouch where gritty materials, such as coarse sand and small pebbles, accumulate for grinding up fibrous feedstuff passing through. Over time, the grit itself gets ground down and must be periodically

Golden Comet hen

G

renewed. **See page 60 for illustration.**
[Also called: ventriculus]

go broody \ Develop the urge to nest, or brood.

Golden Comet \ Trade name for red sex-link hybrid chickens resulting from a cross between a New Hampshire cock and a white Plymouth Rock hen.

Gold Star \ Trade name for red sex-link hybrid chickens.

go light \ Lose weight while eating ravenously, a sign of anemia.

go off \ Give up or stop, as in "to go off feed," meaning to give up eating.

grade \ *See: air cell gauge; egg grading; meat grading*

grain-fed broiler/fryer \ Meat chickens that have been fed up to 70 percent of their diet in scratch grains, resulting in a reduced growth rate and a flavor that is comparable to that of range-fed chickens. Like ranged birds, grain-fed chickens won't be ready for butchering until about 13 weeks. The chickens are fed starter for the first 4 weeks, after which their diet is gradually adjusted to include 70 percent scratch grains, with the remainder being starter (or grower ration, where available) until the time of slaughter.

Where a finisher ration is available, the birds may be fed the starter-grower for their first six weeks, then gradually switched entirely to scratch grains (plus a vitamin/mineral supplement) until the last two weeks before slaughter, when finisher is added to supply 30 percent of the diet. As soon as scratch is introduced, also offer free-choice granite grit to aid in digestion of the grains.

Grand Master Exhibitor \ An award conferred by the American Poultry Association to members who accumulate 100 points by winning a sufficient number of class championships in a single variety of one breed, the point values of which vary with the intensity of the competition.

granite grit \ A form of inert grit made from sifted, crushed granite that is sold in feed stores, sometimes in three grades. *Chick grit* or *fine grit* has the smallest pieces and is suitable for baby chicks. *Grower grit* or *pullet grit* has slightly larger pieces. *Coarse grit,* also called *large grit,* has the largest pieces and is suitable for mature chickens.

Fine sand for chicks and coarse sand for older birds, derived from a clean source, may be substituted for granite grit. Grit collected from roof gutters, where it has washed down from shingles, is unsuitable for chickens, as it contains pollutants from rain, bits of tar from the shingles, molds, and bacteria thriving on the leaves rotting in the gutter, bird and rodent droppings, and other contaminants. *See also: grit*

gravity-flow incubator/ gravity-ventilated incubator \ A mechanical device for hatching eggs that lacks a fan to circulate air. It's called gravity flow, or gravity ventilated, because it relies on gravity to create a natural airflow that allows fresh air to be drawn in through vents in the incubator's bottom as stale, warm air rises and escapes through vents in the top. [*Also called: natural-draft incubator; still-air incubator*]

green egg \ *See: shell color*

grit \ Small pieces of a hard substance eaten by a chicken to aid digestion. Grit lodges in a bird's gizzard, where muscular action breaks down tough substances by grinding them together with the grit. The grit, too, will be ground up and must be replenished.

mineral grit
(oyster shell)

Commercially available grit comes in two forms:

INERT GRIT, such as granite grit, gets ground up rather slowly, in the process furnishing minor amounts of trace minerals.

inert grit
(granite)

MINERAL GRIT, such as oyster shell, gets ground up more quickly and serves as a time-release source of minerals, most notably calcium carbonate (needed by laying hens to produce strong eggshells), and therefore is commonly called calcium grit. Mineral grit should not be used as the sole source of grit for young chickens or mature males, as they can overdose on the calcium.

Chickens that eat only processed feeds — such as chick starter or layer pellets — don't need grit because saliva alone is sufficient to dissolve these feeds. Chickens that eat unprocessed natural feeds need grit to grind up fibrous foodstuffs and make them digestible. Chickens that have access to scratching in dirt generally pick up all the natural

grit they need; chickens in confinement must be fed grit as a free-choice supplement.

ground color \ The main color of a feather's web. Markings such as barring, lacing, penciling, mottling, and spangles appear against the ground color.

ground fed \ Free to peck outdoors, as opposed to being confined to a building or cage.

grower \ A ration for young chickens past the chick stage. \ Young chickens that are still growing.

gullet \ The esophagus, or passageway between the mouth and the crop.

gut flora \ Beneficial microorganisms that live in the intestinal tract. [*Also called: intestinal microflora*]

gynandromorph \ *See: bilateral gynandromorph*

gypsy \ The dark purple, nearly black, color characteristic of the comb, face, and wattles of birchen and brown-red Modern Games, Silkies, Sumatras, and to a lesser extent Sebrights. Gypsy is considered a less-than-preferred color for Japanese bantams and any other breed for which the standard calls for a red comb, face, wattles, and earlobes. [*Also called: mulberry*]

H5N1 \ *See: bird flu*

habituate \ To familiarize. Chickens that have been habituated to humans will remain calm and friendly around people.

hackle \ The collective group of feathers growing along the back and sides of the neck. In a hen these feathers are short and rounded, while in a cock they are longer and pointed, except in hen-feathered breeds. \ Any single one of the hackle feathers.

half sider \ *See: bilateral gynandromorph*

Hall of Fame Exhibitor \ An award conferred by the American Poultry Association to members who accumulate a thousand points as a Master Exhibitor or Grand Master Exhibitor.

Hamburg \ A breed identified in the 1500s as being of Turkish origin, developed in the Netherlands — and known there as Hollands Hoen, or Dutch Chickens — then exported via Hamburg, Germany, to England. There the breed was developed further and identified as the Hamburg. Cold-hardy Hamburgs are good foragers and capable fliers. They have a rose comb, sizable white earlobes, come in a few color varieties, and may be large or bantam. Hamburg hens are excellent layers of white-shell eggs and seldom brood.

See page 138 for illustration.

handpicking \ Manually removing feathers from a chicken being butchered for meat, as opposed to using a picking machine. *See also: dry picking; scalding*

hard feathered \ The tough and tightly held plumage characteristic of the game breeds. The feathers of these breeds are typically short and narrow, have tough shafts, are closely webbed, and have little fluff.

hackle

hackle

hardware cloth \ Sturdy but flexible wire mesh, woven in a grid of squares. Hardware cloth comes in different sizes as measured by the size of the squares: ⅛ (0.3 cm), ¼ (0.6 cm), or ½ (1.3 cm) inch. The larger mesh is often used to make cages or at least cage bottoms, the medium mesh is perfect for screening chicken coop windows to keep out predators and wild birds, and the smaller mesh is used to floor brooders and incubator trays.

hardware disease \ An uncommon condition resulting when the gizzard is pierced from the inside by a sharply pointed object — such as a bit of metal wire, a small shard of glass, or a piece of hard plastic from a broken toy — that the chicken has picked up while pecking on the ground. A chicken with hardware disease will most likely die from infection or starvation. Hardware disease is uncommon in chickens because they don't typically swallow items that cause it.
[Also called: traumatic ventriculitis]

hatch \ The process by which a chick emerges from an incubated egg. \ A group of chicks that come out of their shells at roughly the same time. A normal hatch is completed within 24 hours of the first pip.

hatchability \ The ability of incubated fertile eggs to hatch successfully. Hatchability is affected by the breeder flocks' diet, as well as by exposure of the eggs to filth and extremes of heat or cold prior to incubation. Maximum hatchability is obtained from mature cockerels and pullets that have been laying for at least six weeks. The hatchability of eggs from pullets that have been laying for more than about six months begins to decline — most gradually among bantams and more rapidly among the heavier breeds than among lighter breeds.
As hens get older, the hatchability of their eggs continues a gradual decline — yielding a greater percentage of early embryo deaths and failure of full-term embryos to emerge from the shell — although the chicks that do hatch from eggs laid by older hens will generally be strong, healthy, and disease resistant.

Hamburg hen

Hamburg rooster

hatcher \ A mechanical device, similar to an incubator but with no egg turner, into which eggs are moved from an incubator a few days before they are due to hatch. Using a separate hatcher keeps the incubator more sanitary because contamination by hatching debris — broken shells, chick fluff, and embryo wastes — is confined to the hatcher unit, which after each hatch may be thoroughly cleaned and sanitized before the next hatch.

Using a separate hatcher also offers the advantage of allowing the incubator to continue running at the optimal temperature and humidity for incubation. The hatcher is set for optimal hatching temperature (½ to 1°F cooler [0.3 to 0.5°C] than for incubation) and humidity (6 to 10 percent higher).

hatchery \ A commercial operation that hatches eggs and sells chicks, generally of many different breeds. Hatchery chicks are pretty sure to be healthy and are usually guaranteed for live delivery if shipped by mail. But hatcheries don't offer all the rare and minor breeds or varieties, and since their emphasis is on quantity, the chicks are not often of ideal type, especially for exhibition. Some hatcheries specialize solely in commercial breeds and hybrids developed for exceptional egg or meat production. A large-scale chick seller may not operate a hatchery but may be a broker who sells chicks furnished by specialty breeders or a third-party hatchery.

hatching egg \ An egg destined for incubation, rather than for eating. The primary differences are that hatching eggs must be fertile, while table eggs need not be, and hatching eggs cannot be kept under refrigeration prior to being placed in an incubator.

Only clean eggs should be hatched. Dirty eggs that are extremely valuable may be washed.

Hatching eggs should be of normal color, shape, and size for the breed. The healthiest, most vigorous chicks come from eggs hatched in early spring. *See also: egg storage; egg washing*

hatching position \ The position a chick should assume just prior to hatching to successfully escape from the shell. The chick should have its head under its right wing, with its beak positioned to break through the air cell at the large end of the egg.

hatching
position

hatching record \ A written list of data pertaining to the incubation of eggs. A minimal record includes the number of eggs set, the date they are set, the date they are to hatch, and the number that hatch successfully. Additional information that may prove invaluable includes any problems incurred in regulating the temperature or humidity, the date and length of any power outages, and the condition of any eggs that don't hatch (such as failed to develop, developed but failed to pip, pipped but failed to hatch). The more detailed the record, the more valuable it is for improving future hatching success.

hatchling

hatchling \ A baby chick fresh out of the egg.

hatch rate \ The percentage of fertile eggs that hatch. Assuming the eggs are clean when collected and properly stored prior to incubation and the incubator is run at optimal temperature and humidity, the average hatch rate for artificial incubation is 75 to 85 percent. The hatch rate for natural incubation is higher, usually approaching 100 percent.

head \ The part of a chicken that includes the skull, face, crest (if applicable), comb, eyes, earlobes, beak, and wattles. In breed-standard descriptions, the head is defined as including only the skull and face.

head shaking \ A rapid, jerky movement of the head from side to side. As a feeding behavior, head shaking involves picking up a food item in the beak and shaking it to break off pieces small enough to swallow. As a sign of fear, head shaking involves keeping an eye on the feared object while moving away from it.

health \ The state of being free from disease. To evaluate the health of a chicken, examine the following characteristics:

- **ATTITUDE.** Perky and alert
- **APPEARANCE.** Full, waxy comb; bright eyes; shiny feathers; smooth unsoiled legs
- **ACTIVITY.** Almost constantly pecking, scratching, dusting, preening, or meandering
- **SOUND.** Chattering or singing contentedly

health booster \ *See: supplement*

heart failure \ A condition in which the heart cannot pump enough blood to meet the body's needs. Among chickens, commercial-strain broilers pushed for maximum growth are at the greatest risk for heart failure. The high oxygen demand of rapid muscle growth, combined with lung capillaries that are too small to provide adequate blood flow, results in ascites, a common sign of heart failure. Affected broilers grow more slowly, sit around with ruffled feathers, are reluctant to move, and may die suddenly.

heart rate \ The number of heartbeats per minute. The normal heart rate of a hen at rest is about 312, of a cock is 286. By comparison, the normal heart rate of an adult human at rest is 60 to 100.

heating panel \ A safe source of radiant infrared heat for brooding chicks, nursing a bird back to health, or warming chickens on the roost in extreme winter weather. By directing heat only beneath itself, a panel allows chickens to move away as needed to maintain their comfort level. Sold as pet heaters, these chicken-friendly panels come in several different

heating panel

MINIMIZING HEAT STRESS

- As water consumption goes up, increase the number of watering stations
- Frequently fill drinkers with cool water
- Keep water cool by placing waterers in the shade
- Add electrolytes to water to stimulate drinking and to replenish electrolytes depleted due to the heat
- Avoid medicating the water — if birds don't like the taste, they'll drink less
- Keep rations fresh by purchasing feed more often and in smaller quantities
- Distribute feeders so no chicken has to travel far to eat
- Encourage eating by feeding early in the morning and by turning on lights during cool morning and evening hours
- Open windows and doors and/or install a ceiling fan to increase air movement
- Eliminate crowded conditions by removing some birds or expanding their housing
- Do not confine chickens to hot spaces such as trapnests or cages in direct sunlight or where ventilation is poor
- Provide plenty of shade where the chickens can rest, using a tarp or awning if necessary
- Do not disturb chickens during the heat of the day
- Hose down the coop roof and outside walls several times a day
- Lightly mist adult birds (never chicks; they can chill easily) when the temperature is high and humidity is low

sizes. They are more expensive than infrared heat lamps but use much less electricity and, unlike lightbulb and heat-lamp fixtures, are entirely sealed and therefore easier to clean and sanitize.

heat stress \ Stress induced by a high body temperature that results when the chicken's body produces or absorbs more heat than it can dissipate. Loosely feathered breeds such as Orpingtons and heavily feathered breeds such as the Asiatics (Brahma, Cochin, Langshan) and Americans (Plymouth Rock, New Hampshire, Rhode Island Red) suffer more in hot weather than lightly feathered breeds, and hens in lay suffer more than those not in production.

Mature chickens can adapt to temperature extremes through gradual exposure. A chicken that is acclimated to warm temperatures pants less readily and is less likely to die at what might otherwise be a lethal temperature.

During long periods of extreme heat, hens stop laying, and all chickens suffer stress. When temperatures reach 104°F (40°C) or above, chickens can't lose excess heat fast enough to maintain a proper body temperature and may die. *[Also called: hyperthermia]*

heavy breed \ A breed that tends to be more round than streamlined and in which the hens generally weigh more than 6 pounds (2.7 kg) and the cocks generally weigh 8 pounds (3.6 kg) or more. These breeds were developed primarily for meat production or as utility breeds, although today some of the heavy breeds are largely ornamental.

Hedemora \ An ancient landrace breed originating in the northern part of Sweden and named for the town of Hedemora. It is distinguished by its abundance of down, which keeps this smallish chicken warm in cold weather and makes it look much bigger than it is. Some individuals are entirely silkie feathered (called wool feathered in Sweden).

Hedemoras have short legs, a small single comb, and small wattles. As a landrace they come in a variety of colors, may or may not have slight leg feathering, and may have four toes or five. No breed standard has been established, since the Swedish gene bank (livestock registry) discourages breeding for appearance at the cost of utilitarian value. Hedemoras are active, self sufficient, and extremely cold hardy but cannot tolerate warm climates. The hens lay eggs with light brown shells and are not prolific layers, but they continue laying in the coldest weather. They may brood, and when they do they make excellent mothers.

heirloom breed \ *See: heritage breed*

helminth \ A group of parasitic worms that includes flukes, nematodes, and tapeworms, from the Greek word *helmins*, meaning intestinal worm.

helminthiasis \ The condition of being infested with parasitic worms.

help-out \ A hatching chick that can't get free of the shell without assistance. Help-outs may occur when the hatching humidity is too low, causing shells to stick to the emerging chicks. Help-outs may also occur because either the air cell is in the wrong place or a chick is not properly positioned for hatching. Improper air cell position results from improper turning during incubation or eggs positioned for hatching with the blunt end not higher than the pointed end. An improperly positioned chick — for instance, with its head under its left wing or between its legs — cannot turn to properly to break

Hedemora hen

through the shell. And even when a chick is correctly positioned, if the air cell is not at the large end of the egg, the chick is unable to break free.

When eggs are incubated, then placed in the hatching tray with their large ends upward, and the temperature and humidity is properly controlled, help-outs could be a hereditary issue. Chicks that need help breaking free of the shell, assuming they live, will mature to produce more chicks that have difficulty hatching. For this reason, assisting help-outs is not a good idea.

hen \ A female chicken that is at least one year of age, before which she is a pullet. From the German word *Henne*, meaning hen.

hen feathered \ Having sex feathers, as well as color markings, that are nearly identical to a hen's of the same variety. Hen feathering is a characteristic of Sebrights and to a lesser extent Campines. The latter have a modified form of hen feathering, insofar as the color pattern of same-variety cocks and hens is identical, but the shape of the cock's sex feathers lies between the short, rounded feathers of a hen and the long, pointed feathers of other breeds of cock.

henhouse \ A chicken coop. It is sometimes called a henhouse because in a typical household flock most of the chickens are laying hens, and historically, frugal farmers minimized the feed bill by butchering mature cocks after each spring's hatch.

henmobile \ *See: chicken mobile home*

heritability \ The ability of a characteristic to be genetically transmitted from parents to offspring. Not all inherited characteristics are equally heritable. In general, conformation characteristics are highly heritable, while reproduction characteristics are not. Traits that are highly heritable are easily passed on to offspring, with the result of rapidly improving future generations, while traits with low heritability result in much slower improvement. To perpetuate traits with high heritability, select future breeders on the basis of individual superiority for those traits. To perpetuate traits with low heritability, select breeders on the basis of family averages for the desired traits.

heritage breed \ A traditional breed developed through many years of selective breeding and passed down

EXAMPLES OF HERITABILITY	
Low	High
Egg production	Body weight
Fertility	Growth rate
Chick viability	Immunity to Marek's

through generations, although the exact definition varies with writers and organizations. The Society for the Preservation of Poultry Antiquities defines a heritage breed as being old and rare and defines "old" as existing prior to 1940.

The American Livestock Breeds Conservancy developed a more specific definition for use in labeling meat or eggs sold as being from heritage chickens and suggests that terms such as heirloom, antique, old-fashioned, and old-timey be considered synonymous with heritage. According to this definition, a heritage breed must:

hexagonal
netting

- Have been recognized by the American Poultry Association prior to the mid-twentieth century
- Reproduce through natural mating
- Have the genetic ability to live a long, vigorous life (breeding hens should be productive for five to seven years, cocks for three to five years)
- Thrive outdoors under pasture-based management
- Have a moderate to slow rate of growth (broilers reach market weight in no less than 16 weeks).

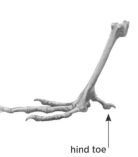

hind toe

heterozygous \ Genes in a pair that differ from each other, from the Greek word *hetero*, meaning different, and *zygous*, meaning pair. \ An individual with a large number of paired dissimilar genes. When heterozygous chickens are mated or an outcross is made, the genes in their offspring can pair in many different combinations, with unpredictable results in the offspring.

hexagonal netting/wire \
Wire mesh consisting of thin wire twisted and woven together into a series of hexagons, giving it a honeycomb appearance. Mesh sizes range from 0.5 inch to 2 inches (1.5 to 5 cm). The smallest mesh is generally used for chicks. The 1-inch (2.5 cm) mesh is more commonly used for mature chickens than the largest, which is difficult to stretch properly tight. This lightweight fencing will confine chickens, but it tears easily, and determined predators will rip the mesh to break in. The best use for hexagonal netting is as dividers for indoor pens or as a cover for an outdoor run to keep out flying predators.
[Also called: chicken wire]

hex net/wire \ Short for hexagonal netting/wire.

hind toe \ The fourth or back toe, which should point straight to the rear and lie firmly on the ground to give the chicken proper balance. *See also: anatomy*

hock/hock joint \ The joint between a chicken's lower thigh and shank, corresponding to the human ankle. *See also: anatomy, skeletal*

holding funnel \ *See: killing cone*

Holland \ A breed developed in the United States as a dual-purpose farmstead chicken that produces white-shell eggs. The Holland was created by crossing several different breeds, including some chickens imported from the Netherlands; hence the name.

The Holland has a single comb, may be large or bantam, and comes in two color varieties — barred and white, of which the barred has always been more popular and the white in fact may no longer exist. The hens lay white-shell eggs and are not particularly inclined toward broodiness.

hollow comb \ An undesirable depression at the top of a cushion comb, pea comb, rose comb, strawberry comb, or walnut comb.

homozygous \ Paired genes that are identical to each other, from the Greek word *homo*, meaning same, and *zygous*, meaning pair. \ An individual with a large number of paired identical genes. The more closely chickens are related, or inbred, the more homozygous they become and the more predictable their offspring.

hormones \ Synthetic estrogens used initially in the 1950s to increase the weight gain of chickens but discontinued in the late 1970s. Federal regulations now ban the use of hormones for poultry, yet sellers of eggs, chicken meat, and chicken feed sometimes label their products as "hormone free" to imply superiority to other brands.

horn \ A beak color consisting of light and shaded areas, as is characteristic of blue Andalusians, Rhode Island Reds, speckled Sussex, and others.

horn comb \ *See: V-comb*

horny \ Description of a hornlike material, such as that making up a chicken's beak, the scales on its legs, its spurs, and its toenails.

host \ A bird (or other animal) on or in which a parasite or an infectious agent lives.

Houdan \ An old French breed, named after the town of Houdan, which traditionally has been a source of poultry for the Paris market. The Houdan is a dual-purpose chicken that is also ornamental. Its crest, V-comb, and beard likely were bestowed by Crevecoeur and Polish ancestors, while its five toes were contributed by the Dorking. The Houdan may be either large or bantam in one of two color varieties — mottled is the original; white was developed in the United States. The hens lay white-shell

Holland hen

eggs and tend toward broodiness but are not particularly adept at it.

house chicken \ A chicken kept as a house pet, preferably of a docile breed. Before acquiring a house chicken, it's a good idea to make sure no one living in the household is allergic to chicken dander. The easiest way to do that is for each family member to spend time visiting someone else's chickens before you make a commitment to bring one into the house.

Even a house chicken needs to spend daily time outdoors engaging in normal chicken behaviors — sunbathing and dust bathing, scratching in dirt, and snacking on bugs and green leaves. The biggest challenge to keeping a house chicken is that it cannot easily be housebroken, which is why the chicken diaper was invented.

Houdan hen

hover \ A heat source hung above brooded chicks so it hovers over them when they huddle beneath it. A hover may be gas or electric, round or rectangular, and with or without curtains hanging around the edges to keep in heat and keep out drafts. A hover is designed to allow chicks to warm themselves as needed or to move away from the heat to eat, drink, and exercise.

humane treatment \ Marked by kindness and compassion. Claims on meat or egg labels are not always regulated and therefore may not be true unless they are certified, meaning a certifying agency verifies the chickens are handled gently to minimize stress, are given ample fresh water and proper nutrition, and have sufficient living space to engage in natural behavior.

hutch \ A cage, usually with a wire-mesh front and elevated off the ground by being either mounted on legs or suspended from a wall or ceiling.

hybrid \ The offspring of a rooster and a hen of different breeds, each of which might themselves be hybrids. Commercial broilers and brown-egg layers are hybrids. In designating a hybrid the male's breed or strain is named first. For example, Cornish Rock is a cross between a Cornish rooster and a Plymouth Rock hen. This cross results in fast-growing broilers raised in confinement, while the Freedom Ranger and similar crossbreeds are more suited to pasture broiler production. Black sex-links, red sex-links, and production reds are hybrid brown-egg layers. Commercial white-egg layers are not hybrids in the strictest sense; they are bred from different strains of the same breed and variety — single-comb white Leghorn. Since hybrids result from matings between different breeds (or highly specialized strains within a single breed), they won't breed true.   *See also: crossbreed*

hybrid vigor \ The phenomenon whereby crossbred chicks are stronger and healthier than either of their parents. Hybrid vigor results from maintaining a high degree of heterozygosity in the breeder flock by continually outcrossing or semioutcrossing. Hybrid vigor is the opposite of inbreeding depression. Traits with low heritability that show the greatest degree of inbreeding depression — such as reproductive performance and chick viability — respond the strongest to hybrid vigor.

hygrometer \ *See: wet-bulb thermometer*

hyperthermia \ *See: heat stress*

hypnotism \ The practice of putting a chicken into a trance wherein it apparently loses its ability to move. Hypnotism slightly reduces a chicken's heart and respiration rate, helping calm a nervous bird. Although the hypnotized chicken is unharmed, once let go it won't move for several seconds, sometimes as much as a minute.

HOW TO HYPNOTIZE A CHICKEN

Some methods for hypnotizing a chicken, which may need to be repeated a time or two to be effective, are:

- Turn the chicken onto its back, hold it with one hand, and with the other hand gently stroke its throat
- Turn the chicken onto its back, hold it with one hand, and with the thumb and index finger of the other hand, gently stroke both sides of its breast
- Turn the chicken onto its back and slowly run a finger from its wattles down almost to the vent
- Place the chicken's head under one of its wings, and gently rock the chicken back and forth; then carefully set it on the ground
- Lay the chicken on its side with its head on the ground, and hold it firmly for about 30 seconds before letting go
- Lay the chicken on its side with its legs stretched back and its head on the ground, point a stick or your finger just in front of its beak (taking care not to touch the beak), and repeatedly and quickly draw a 4-inch (10 cm) line on the ground straight out and parallel to the beak
- With the chicken standing, repeatedly draw your finger along the ground straight out from its under its beak to about 6 inches (15 cm) in front of the chicken

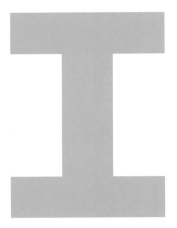

immunity \ Ability to resist infection. Active immunity confers resistance to a disease as a result of having had the disease or having been vaccinated against it. Passive immunity results when the body manufactures antibodies that combat a disease as a result of being injected with an antiserum.

impacted crop \ *See: crop impaction*

impaction \ Blockage of a body passage, such as the crop, as a result of eating something the chicken can't digest, or the cloaca, as a result of a stuck egg.

inbred \ Produced by inbreeding.

inbreeding \ Mating closely related individuals, especially over several generations. Inbreeding has the effect of concentrating a specific set of genes, and in doing so reveals recessive traits, which can be a good thing or a bad thing. If a recessive is desirable, you want to encourage it. If it is undesirable, you want to weed it out, which is possible only by maintaining sufficient genetic diversity to prevent the concentration of undesirable recessives in all the breeders.

inbreeding depression \ Decrease in vigor and fitness in the offspring of inbred parents. By concentrating genes, inbreeding creates uniformity of size, color, and type but also brings out weaknesses, such as reduced rate of lay, low fertility, poor hatchability, and slow growth. Inbreeding doesn't cause these problems but accentuates any tendency toward them.

Some strains are less susceptible than others to the effects of inbreeding depression. Popular breeds with lots of varieties offer more opportunities for avoiding inbreeding depression than less common breeds and those with few varieties.

To minimize inbreeding depression, avoid brother-sister and offspring-parent matings and instead mate birds to their grandsires or granddams. Retain as future breeders those with the best fertility, hatchability, chick viability, disease resistance, and body size. Never breed birds that lean toward infertility. By inbreeding gradually and choosing breeder cocks and hens carefully, such traits as laying ability and disease resistance can be improved.

Red flags indicating an outcross is needed include the following:

- The unexpected appearance of an undesirable trait
- A rapid or drastic reduction in fertility, hatchability, chick viability, or general health
- A continuing lack of improvement, indicating the breeder flock lacks the right genes

incubate \ To maintain adequate warmth, humidity, and other favorable conditions for the hatching of fertile eggs.

incubation humidity \ The amount of moisture in the air inside an incubator. For a successful hatch, moisture must evaporate from the eggs at just the right rate, and evaporation is regulated by the amount of moisture in the air — the more moisture-laden the air, the more slowly moisture evaporates from the eggs and vice versa. Overly rapid evaporation (too-low incubation humidity) can inhibit the chicks' ability to get out of their shells at hatching time. Overly slow evaporation (too-high humidity) can lead to mushy chick disease (omphalitis). To slow the rate of evaporation from the eggs, every incubator has a water-holding device that gradually releases moisture into the air, and many incubators have a method of controlling the rate of evaporation from the water-holding device by adjusting the surface area available for evaporation.

An accumulation of moisture on the incubator's observation window during the hatch indicates excessive humidity. To release excess humidity some incubators have additional vents to those needed for good airflow. These extra vents are closed in some way, such as with a plug or sliding cover. Opening, or partially opening, some or all of these vents decreases humidity by allowing more moisture-laden air to escape.

HUMIDITY AT COMMON INCUBATION TEMPERATURES

Circulated Air		Gravity Flow						Relative Humidity
99.5°F/37.5°C TEMPERATURE		**100°F/37.8°C** TEMPERATURE		**101°F/38.3°C** TEMPERATURE		**102°F/38.9°C** TEMPERATURE		
Wet-Bulb Reading								
80.8°F	27.1°C	81.3°F	27.4°C	82.2°F	27.9°C	83.0°F	28.3°C	45%
82.8	28.2	83.3	28.5	84.2	29.0	85.0	29.5	50%
84.7	29.3	85.3	29.6	86.2	30.1	87.0	30.6	55%
86.7	30.4	87.3	30.7	88.2	31.2	89.0	31.7	60%
88.5	31.4	89.0	31.7	90.0	32.2	91.0	32.8	65%
90.3	32.4	90.7	32.6	91.7	33.2	92.7	33.7	70%
91.9	33.3	92.5	33.6	93.6	34.2	94.5	34.7	75%
93.6	34.2	94.1	34.5	95.2	35.1	96.1	35.6	80%
95.2	35.1	95.7	35.4	96.8	36.0	97.7	36.5	85%

From: *Storey's Guide to Raising Chickens,* 3rd edition

Low humidity is a more common problem. If humidity drops inexplicably during a hatch, the water pan's surface may be coated with fluff released by newly hatched chicks, preventing evaporation and causing the humidity to plummet. Since the eggs contribute to humidity by evaporating through the shell, an incubator that is not filled to capacity may have more difficulty maintaining proper humidity than a full incubator. The larger the incubator and the less full it is, the greater the problem. If eggs hatch in exactly 21 days but not all pips hatch and the hatch rate of fertile eggs is generally poor, most likely improper humidity is at fault.

Most incubators call for about 60 percent relative humidity, except during the last three days prior to the hatch, when it should be increased to 70 percent. An electronic hygrometer, or an incubator with an electronic control system, shows the humidity level in a digital display. Humidity in a nonelectronic incubator is measured by a wet-bulb thermometer. In place of a hygrometer, or in conjunction with one, a good indication of humidity is the changing air-cell size inside the developing eggs, as determined by candling.   *See also: air cell*

incubation period \ The time

needed for an egg to hatch, which for chickens is 21 days. \ The time elapsed between exposure to a disease-causing agent and the appearance of the first signs. The incubation period varies from disease to disease, and may be a matter of hours or weeks.

incubation temperature \

The typical operating temperature for a circulated-air incubator is 99.5°F (37.5°C); for a still-air incubator, 102°F (38.9°C). Lethal temperatures are 103°F (39.5°C) in a circulated-air incubator, 107°F (41.7°C) in a still-air incubator. The recommendations given in the operating manual should be considered baseline; you can improve your success by making minor adjustments based on past hatching records.

At an incubation temperature that is ½ to 1°F (0.3 to 0.5°C) lower than optimal, chicks will take longer than the normal 21 days to hatch. They will tend to be big and soft with unhealed navels, crooked toes, and thin legs. They may develop slowly or may never learn to eat and drink and therefore will die. If the temperature runs ½ to 1°F higher than optimal, chicks will hatch before 21 days. They will tend to have splayed legs and can't walk properly.

To keep the temperature steady, position the incubator away from drafts and where the room temperature remains fairly constant. A room temperature between 75 and 80°F (24 and 27°C) is ideal, although between 55 and 90°F (13 and 32°C) is acceptable. Occasional minor fluctuations are normal, so once the incubator's temperature is properly set, resist the temptation to keep adjusting it.

incubator/hatcher sanitation \ Scrubbing out an incu-

bator between hatches. Good incubator sanitation improves hatching success, gives chicks a healthy start in life, and helps break the natural disease cycle in a flock. Because hatching itself is a major source of contamination, thoroughly clean out the incubator or hatcher between hatches and at the end of every hatching season.

With no eggs in the incubator or hatcher, unplug the unit, remove and clean all removable parts, vacuum up loose fluff, sponge out hatching debris, and scrub the incubator with detergent and hot water. Take care not to wet down or spray the heater or any electrical parts — instead, brush them off with a soft-bristle paintbrush and gently move a vacuum hose nearby to pick up the loosened fluff. When the incubator and all its parts are clean of debris, wipe or spray the nonelectrical parts with a sanitizer, then leave the incubator open until it is thoroughly dry, preferably in sunlight, before storing it in a clean place.

incubator ventilation \ A provision for circulating fresh air inside an incubator. Developing embryos use up oxygen rather rapidly, while at the same time generating carbon dioxide. An incubator therefore needs good airflow to constantly replenish oxygen and remove carbon dioxide. For this purpose, all incubators have vents that must remain open at all times for adequate oxygen intake. Some incubators have additional covered vents that may be opened to increase airflow when necessary to reduce excess humidity.

Indian Game \ British term for the Cornish breed.

industrial \ *See: breed*

INCUBATOR BASICS

Incubators come in a broad range of sizes from small ones that hold only three eggs to large room-sized commercial incubators that hatch thousands of eggs at a time. Practical incubators for home use hold anywhere from a few dozen to a few hundred eggs. Small incubators are generally tabletop models with a lid at the top, and may be either forced-air or still-air operated. A large incubator is always forced air; it is built like a cabinet that opens at the side and has pullout trays to hold the eggs. Both tabletop and cabinet units range from hands-on models that require frequent attention to digital units that handle everything electronically. The five important features of any incubator are:

- Method of turning
- Airflow control
- Temperature control
- Humidity control
- Ease of cleaning

See also: egg turning; incubation humidity; incubation temperature; incubator/hatcher sanitation; incubator ventilation

inert grit \ A hard form of grit, such as granite grit or washed river sand, that does not readily get ground up by the gizzard, as compared to a mineral grit, such as oyster shell, that gets ground up more quickly and in the process releases appreciable amounts of dietary calcium and other minerals. Grit fed to young chickens, and to mature males, should be inert, as they can overdose on calcium. Laying hens should have access to inert grit as well as mineral grit, so they can adjust their intake as needed.

infectious \ Capable of invading living tissue and multiplying therein, causing disease.

infectious anemia \ A viral disease of the blood occurring mainly in broilers raised on reused litter. The primary sign is sudden deaths among apparently healthy chickens. Some affected chickens develop diarrhea, which may be bloody, causing infectious anemia to be mistaken for coccidiosis.
[Also called: chicken anemia]   See also: anemia

infectious bronchitis (IB) \ One of several viruses that cause cold-like signs, including coughing, runny nose, and swollen eyes. It is often spread at poultry shows and is so contagious it can be transmitted through the air from one chicken to another 1,000 yards (915 m) away. It causes few deaths; infected chickens generally recover within two to three weeks, but they remain carriers.

infectious laryngotracheitis \ A highly infectious viral disease of the upper respiratory tract that is often transmitted at poultry shows. It causes similar cold-like signs to those of infectious bronchitis but spreads less rapidly and is more severe. Most infected chickens either die or recover within two weeks. Survivors are carriers, and therefore, infectious laryngotracheitis is reportable in some states.
[Also called: laryngo]

infertile \ Describes an egg that has not been inseminated and therefore cannot hatch, most commonly because no cock is present. \ A cock that is temporarily or permanently unable to produce sperm, or a hen that is temporarily or permanently unable to lay eggs.

infertility \ Temporary or permanent inability to reproduce, the reasons for which are many and varied and often hard to ascertain. Causes of infertility include:

- Inadequate nutrition
- Obesity
- Reaction to a drug or toxin, especially aflatoxicosis
- Poor condition
- Illness
- Stress
- Age (too young or too old)
- Temperature extremes
- Inbreeding depression
- Too many cocks (that spend more time fighting than mating) or too few cocks (that can't get around to all the hens)

Season also affects fertility, which tends to be lowest when daylight hours are fewer than 14, as well as during the heat of summer.

initial vaccination \ The first vaccination in a series, the remainder of which are called boosters.

inner thick \ The layer of dense albumen immediately surrounding the yolk of an egg. It is called the inner thick to distinguish it from the other thick layer of egg white, which is called the outer thick.
[Also called: chalaziferous layer]

insecticide \ A product used to minimize intermediate hosts in or around a chicken coop or to eliminate external parasites from the chickens' bodies. The list of insecticides approved for poultry is short and changes often. State Extension poultry specialists and veterinarians have the latest information. Never use a nonapproved product on chickens, especially those raised for meat or eggs. Even an approved insecticide is toxic and must be handled with care; read labels and follow all precautions.

intensity of lay \ The number of eggs a hen lays during a given time.

intermediate host \ An organism that supports the immature or nonreproductive forms of a parasite. Most roundworms and all tapeworms require an intermediate host, meaning a chicken cannot become infected by eating such a parasite's egg coming directly from an infected chicken but must eat an intermediate host containing a parasite egg. *See also: life cycle, indirect*

intestinal microflora \ Beneficial microorganisms that live in the intestinal tract.
[Also called: gut flora]

inverted spike \ An undesirable feature of a rose comb in which the rear portion appears to be indented, or telescoped, into the back of the comb instead of extending outward.

PARASITIC WORMS AND THEIR HOSTS

Parasite	Intermediate Host
Ascarid	None (direct cycle)
Capillary	None or earthworm
Cecal	None or beetle, earwig, grasshopper
Fluke	Dragonfly, mayfly
Gapeworm	None or earthworm, slug, snail
Tapeworm	Ant, beetle, earthworm, grasshopper housefly, slug, snail, termite

From: *Storey's Guide to Raising Chickens,* 3rd edition

Japanese Bantam \ An ancient breed of true bantam developed in Japan. The Japanese Bantam has a large tail in proportion to its size and carries the tail so far forward that the cock's tail feathers

Japanese Bantam
rooster

nearly touch the back of its head. These little birds are capable fliers and come in several color varieties, as well as in bearded and nonbearded varieties. The Japanese has a large single comb, long drooping wings, and short legs. The short legs are associated with the lethal creeper gene. The hens lay brown-shell eggs and make excellent broodies.

Japanese Silkie \ *See: Silkie*

Java \ A dual-purpose farmstead breed brought to the United States and developed by early settlers; now quite rare. It was named after the Indonesian island of Java, although no one is certain that's where the breed originated. The Java was used in developing the Jersey Giant and the Plymouth Rock, both of which eventually overshadowed the Java in popularity. The breed may be identified by its extremely long sloping back and its five-pointed single comb, with the first point appearing farther back on the head compared to most other single-comb breeds. The Java may be large or bantam and comes in four color varieties — auburn, black, mottled, and white. The hens lay eggs with brown shells and make excellent broodies.

Jersey Giant \ A breed developed in the United States and named after New Jersey, its state of origin. The Jersey Giant was created as an alternative to turkey and is the largest known breed — hens mature to 10 pounds (4.5 kg), cocks to 13 pounds (6 kg), and caponized cockerels reach a whopping 20 pounds (9 kg). Raising Jersey Giants for meat production turned out to be uneconomical because they put most of their energy into developing a strong bone structure before they start fleshing out at about six months, and take as long as nine months to reach a reasonable size for harvesting. They remain popular, however, for their huge size, calm disposition, and cold hardiness. The Jersey Giant has a single comb, may be large or bantam, and comes in a few color varieties, of which black and white are the two most common. For such a large breed, the hens are good layers of brown-shell eggs and make wonderful broodies, although their eggs may take a day or two longer to hatch than other breeds.

joint ill \ *See: arthritis*

judging system \ The system according to which chickens are evaluated at an exhibition. In the United States, one of three systems is used:

THE AMERICAN SYSTEM is generally used for adult and open shows. The chickens in each class are ranked against each other according to the standard for their breed and variety. Awards are generally given to the top three or five, although some shows go all the way down the line.

THE DANISH SYSTEM is often used for youth shows. The chickens are not compared to one another but rather each bird is judged on its own merit according to how well it meets the standard description for its breed and variety. Instead of pitting competitors against one another, this system helps exhibitors gauge their individual progress.

Jersey Gaint hen

Java hen

A MODIFIED DANISH SYSTEM is used by 4-H. To avoid discouraging younger members and those new to poultry, the standard is adjusted according to each exhibitor's age and years of experience.

jungle fowl \ Four species of game bird originating in southern Asian, and typically living in forested areas, that are related to the domestic chicken. Of the four species — the green jungle fowl (*Gallus varius*); the red jungle fowl (*G. gallus*); the gray jungle fowl (*G. sonneratii*); and the Sri Lankan jungle fowl (*G. lafayetii*), also known as Ceylon jungle fowl — the red is probably the primary ancestor of all domestic chickens, although other species, notably grays, were likely involved. The red jungle fowl is distinct from most modern chickens in being more gamelike in appearance and carrying its tail horizontally, and the hen has only a rudimentary comb.

red jungle fowl

keel \ The ridge running down the outer center of a chicken's breastbone, resembling the keel of a boat. Chickens and other birds are the only vertebrates having a keel, which serves as an attachment for wing muscles.

keel bone \ *See: breastbone*

keel bursitis \ *See: breast blister*

keel cyst \ *See: breast blister*

kelp meal \ A feed supplement derived by drying and grinding up brown cold-water seaweeds of the orders Laminariales and Fucales. Kelp meal is rich in a large number of vitamins, amino acids, and especially trace minerals. It is used as a free-choice supplement for layers, to strengthen eggshells and darken the yolks; for breeders, to increase fertility and vitality; and for show chickens, to improve condition and enhance immunity.

keratin sheath \ The thin, fibrous tubelike structure covering a newly emerging feather when a chicken molts. The keratin sheaths come off as the chicken preens or scratches itself with a claw or, in the case of hard-to-reach parts of the anatomy, eventually fall off on their own.

kerosene incubator \ An incubator fueled by kerosene as a source of heat. Kerosene incubators are used by off-grid poultry keepers and those living where power outages are frequent. Getting a constant temperature requires trial and error, and the best hatch rate is about 75 percent.

Kestin score \ A method of assessing the ability of a chicken to walk based on a six-point scale as follows:

0 No sign of lameness
1 A slight but detectable lameness
2 An abnormal gait that does not affect the chicken's ability to find food and water
3 The chicken is so severely impaired it has trouble getting to food and water
4 The chicken can stand up but doesn't want to walk
5 The chicken is so lame it cannot stand up

key feather \ *See: axial feather*

killing cone \ A funnel-shaped device into which a chicken is inserted, head first, to keep it still for slaughter. Stainless steel killing cones are available in various sizes to fit the size of the chicken snugly; a traffic cone, or an empty bleach or laundry detergent jug

knob

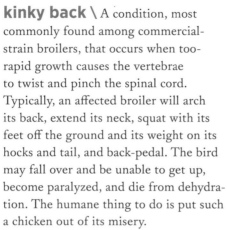

with the top and bottom cut off, makes a workable substitute.

[Also called: holding funnel]

king's comb or crest \

See: carnation comb

kinky back \ A condition, most

commonly found among commercial-strain broilers, that occurs when too-rapid growth causes the vertebrae to twist and pinch the spinal cord. Typically, an affected broiler will arch its back, extend its neck, squat with its feet off the ground and its weight on its hocks and tail, and back-pedal. The bird may fall over and be unable to get up, become paralyzed, and die from dehydration. The humane thing to do is put such a chicken out of its misery.

knee/knee joint \ The thigh joint,

located between the upper thigh and the lower thigh.

See page 18 for illustration.

knob \ A lump on top of the head of

a crested chicken, from which the crest feathers grow. The knob is part of the skull and consists of bone, tissue, and feather follicles.

knock-knee \ A leg deformity in

which the hocks are closer together than the feet.

Kraienkoppe \ A breed originating

in the Netherlands as a fighting chicken and developed in Germany as an ornamental. The origin of the breed name is unknown and most likely is a German dialect. *Krähen* is the German word for crow, perhaps in reference to the similarity of the shape of this chicken's head to that of a crow. The Kraienkoppe has the body conformation of a Leghorn, one of its ancestors, with the tight feathering and fierce expression of a fighting breed (Malay is another ancestor). It has a small walnut comb and a beetle brow, comes in a few colors (not all of which are found in North America), and may be large or bantam (although the bantam size is not found in North America).

These chickens are capable fliers, excellent foragers, cold hardy, and disease resistant. The hens lay eggs with slightly tinted shells. The breed is still quite rare in North America and differs from the same breed in Germany in that the North American Kraienkoppe is smaller; the hens don't lay quite as well; and the hens brood more readily.

Kraienkoppe hen

K

Label Rouge

Label Rouge \ French certification program (meaning Red Label) for naturally raised poultry, the feeding method of which has become a model for American organic broiler growers. The Label Rouge ration is of 100 percent vegetable origin, supplemented by insects and plants the chickens find in the pine forest where they actively forage. The main element in the diet of the French broilers is corn, crushed and mixed into a ration consisting of:

80% CORN (FOR ENERGY)
15% SOY (FOR PROTEIN)
3% MINERALS AND VITAMINS
2% ALFALFA

Label Rouge broilers are not of the Cornish-cross commercial type but are a cross between the slower-growing heritage Cornish and other old-time breeds. These broilers do not have white plumage like that of industrial broilers; hence, some of the trade names for similar hybrids in the United States include a reference to color: Black Broiler, Color Yield, Colored Range, Freedom Ranger, Kosher King, Red Broiler, Redbro, Red Meat Maker, Redpac, Rosambro, and Silver Cross, to name a few.

laced, lacing \ A narrow border of contrasting color of uniform width that extends all the way around the edges of a feather.

La Fleche \ An old meat breed from France, named after the village of La Flèche. This large chicken has a V-comb similar to the comb of several crested breeds, and indeed this chicken bears a vestigial crest. The La Fleche has white earlobes, long wattles, a broad breast, a full tail, and tight feathering. It may be large or bantam and most commonly has black plumage, although other color varieties have been bred.

As a breed these chickens are wary, somewhat feral, capable fliers, and

laced

La Fleche rooster

La Fleche hen

for such rapid growth that their heavy bodies put a great strain on their legs and joints, and they can't get around well. A small percentage can't walk at all and therefore can't get to feed and water, leaving them to be trampled by the more mobile birds and eventually to die of starvation or dehydration.

Lameness scoring systems have been devised to determine when a review of management practices might be needed and at what point a lame bird should be humanely put down. A six-point system, called a Kestin score, was developed in England and simplified into the U.S. three-point gait-scoring system.

self-sufficient but not winter hardy. The hens lay white-shell eggs and seldom brood. *See also: devil bird*

Lakenvelder \ An ancient laying breed originating in the Netherlands and named after the village of Lakenvelt. The same region has similarly two-toned cattle with the same breed name, known in North America as Dutch Belted.

The Lakenvelder chicken is small, has a single comb, may be large or bantam, and comes in a few colors, of which the most common is white with a black head, neck, and tail.

Lakenvelders are active, wary, and somewhat flighty. The hens are good layers of eggs with tinted shells and seldom brood.

lameness \ Impaired ability to walk. Lameness in breeder cocks can be a problem, since a lame rooster has trouble hanging onto a hen while breeding. The greater issue is among commercial-strain Cornish-cross broilers, developed

Lakenvelder rooster

Lamona \ A dual-purpose breed developed in the United States as a chicken with white plumage that produces eggs with white shells and lays as well as a Leghorn but is meatier. The breed was named after Harry S. Lamon, the man who developed it in the 1920s. By the 1950s the Cornish Rock cross took over the meat industry, the white Leghorn took over the egg industry, and the Lamona gradually disappeared. Since the 1980s no source of stock has been identified.

landrace \ A population of chickens (or other livestock, or plants) that has been developed over many centuries through natural selection and the effort of breeders to select for characteristics best suited to local needs and the local environment. The resulting chickens are genetically consistent enough to be considered a breed but are not as consistently uniform in appearance as a breed that has been standardized. Typically, a landrace may be fairly uniform in size, laying ability, and hardiness but may not be uniform in plumage color or eggshell color or in the appearance of such features as beards, crests, or feathered legs. *See also: breed*

Langshan \ An old dual-purpose breed from China, named after the town of Langshan at the foot of the Langshan mountain range in northern China. The Langshan is one of the tallest breeds, given a stately appearance by its full tail,

Lakenvelder hen

which it carries high, and its long, lightly feathered legs. It has a single comb, comes in a few colors, of which black is the most common, and may be large or bantam.

This hardy, tight-feathered breed adapts well to warm or cold climates and is a capable flier for a chicken of its size. The hens are good layers of dark-brown eggs and will brood, though not always successfully.   *See also: Croad Langshan*

large roundworms \ *See: ascarids*

laryngo/laryngotracheitis \
See: infectious laryngotracheitis

lavender \ *See: blue*

laxative flush \ A method of cleansing the system of a chicken that is suffering from poisoning or an intestinal disease. The laxative flush hastens recovery by absorbing toxins and removing them from the body. A solution of Epsom salts (magnesium sulfate) makes the best flush, but chickens don't like the taste and won't readily drink it, so they must be treated individually. If a number of birds are involved, or handling them would cause undue stress, molasses may be added to the drinking water to flush the whole flock at once. Either way, flush only adult chickens, never chicks.

Langshan hen

L

Langshan rooster

LAXATIVE FLUSH RECIPES

EPSOM SALTS FLUSH: 1 teaspoon (5 mL) Epsom salts in ½ cup (118 mL) water, squirted down the bird's throat twice daily for two to three days, or until the bird recovers.

MOLASSES FLUSH: 1 pint (0.5 L) molasses per 5 gallons (19 L) water, left in the drinker for no longer than eight hours, then replaced with fresh, clean water.

layer \ A hen that is currently laying eggs. Most hens lay best during their first year, although a really good layer should do well for two years or even three. To tell the difference between a good layer and a poor layer, consider the following features:

- A high producer is active and alert. A low producer tends to be lazy and listless.
- The feathers of a good layer are worn, dirty, and broken. Poor layers look sleek and shiny.
- The skin of a good layer is stretchy and bleached out. The skin of a poor layer is tight and in a yellow-skin breed retains full color.
- The comb and wattles of a good layer are large, bright, and waxy. Candidates for culling have small combs and wattles.
- The legs of a good layer are wide apart and set back, and the shanks are thin and flattened at the sides. A poor layer has legs that are set forward and close together — indicating less body capacity for egg-producing organs — and the shanks are round and full.

- The vent of a good layer is large, moist, and oval. Cull candidates have tight, dry, round vents.
- A good layer's abdomen should feel soft, round, and pliable under your hand — never small and hard. But be careful not to mistake an about-to-be-laid egg for a hard abdomen.
- Pubic bones should have enough room between them for three or more fingers for most breeds, at least two fingers for small breeds. The distance between the pubic bones and keel should accommodate at least four fingers; the greater the distance, the better the layer. A lazy layer is tight and nonflexible in these two areas.

Of all these various indicators, the most reliable ones are a moist vent and flexible pubic bones. A hen with a puckered vent and stiff or thick, inflexible pubic bones is not laying, period.

laying breed \ A breed generally known to lay nearly one egg per day for long periods at a time. Other breeds lay fewer eggs either because they take longer rest periods between bouts of laying or because they tend to go broody. Some breeds may be just as prolific as the typical layer breeds but eat more feed per dozen eggs and therefore do not make economical layers. Efficient laying hens share four desirable characteristics:

- They lay a large numbers of eggs per year. The best layers average between 250 and 280 eggs per year, although individual birds may exceed 300.
- They have small bodies. Compared to larger hens, small-bodied birds need less feed to maintain adequate muscle mass.
- They begin laying at 17 to 21 weeks of age. Dual-purpose hens, by comparison, generally start laying at 24 to 26 weeks.
- They do not typically get broody. Since a hen stops laying once she begins to nest, the best layers don't readily brood.

The most efficient laying breeds tend to be nervous or flighty. But kept in small numbers in uncrowded conditions, with care to avoid stress, these breeds can work fine in a backyard setting.
See also: rate of lay

laying cycle \ The number of consecutive days during which a hen lays without a break, which can vary in length from 12 days to nearly a year. The best heavy-breed hens in peak production lay about 40 eggs in a cycle; a typical Leghorn lays closer to 80.

lay ration/layer ration \ *See: feeding hens; feeding pullets*

leader \ *See: spike*

LAYING BREEDS (NONHYBRID)

Breed	Rate of Lay
Ameraucana	Good
Ancona	Best
Andalusian	Good
Araucana	Good
Australorp	Good
Barnevelder	Good
Campine	Good
Catalana	Good
Chantecler	Good
Dominique	Good
Empordanesa	Good
Fayoumi	Best
Hamburg	Good
Lakenvelder	Good
Leghorn	Best
Marans	Good
Minorca	Best
Norwegian Jaerhon	Best
Penedesenca	Good
Plymouth Rock	Good
Rhode Island Red	Good
Rhode Island White	Good
Welsumer	Good

Good = 150–200 per year; best = close to 300. Blue type indicates better layers among the dual-purpose breeds.

leaker \ An egg with a cracked shell and a broken shell membrane, which allows the contents to ooze out. A leaker is unsafe to eat.

leg \ In common usage a chicken's leg includes the upper and lower thigh and the shank. However, anatomically, the shank is part of the foot, making the hock joint the chicken's ankle, so technically, the leg consists of just the upper and lower thigh.

leg band \ A plastic or aluminum ring that wraps around a chicken's leg for the purpose of identification. Leg bands come in different sizes and colors, with numbers (called bands or bandettes) or without (called spiral bands or spirals). A leg band must be the right size for the bird to ensure it is neither so small it binds nor so big it falls off.

Different breeds may require different sizes, and within a breed the cocks usually require a larger size than the hens. Younger chickens need a smaller size than mature chickens and may require several size changes before they are fully grown. When banding growing birds, check frequently and replace any band that gets too tight to slide up and down easily. As spurs grow make sure the band remains above the spur, where it won't bind. A band that becomes imbedded in a leg can cause lameness.

Leg-band sizes range from #2 to #16, not all of which are suitable for use with chickens. Most bantams fall between #5 and #9, and most large breeds fall between #9 and #12. The size denotes the band's inside diameter in 16ths of an inch. The American Bantam Association uses its own sizing system in millimeters, with corresponding letter designations, as follows: D-10 (equivalent to #6), E-11 (#7), F-13 (#8), G-15 (#9), H-18 (#11), and I-20 (#13).

leg

upper thigh

lower thigh

shank

leg color \ The color of a chicken's shank, the most common colors being white, yellow, and gray. Leg color varies from breed to breed, and some breeds have different leg colors for different varieties.

Leghorn \ A breed developed in Italy and named after the port city of Livorno (called Leghorn in English). Leghorns may be large or bantam and come in many color varieties and two comb varieties: single comb and rose comb. This small chicken with a full tail has a reputation for being noisy, nervous, and flighty. Leghorns remain popular they

Leghorn hen

legband

L

are early maturing, hardy, heat tolerant, and fantastic layers, and they have good fertility and superior feed-conversion efficiency. Because of these reasons, the Leghorn is the breed of choice for the egg industry, which has developed a limited number of highly specialized single-comb strains. The hens are prolific layers of large eggs with white shells and seldom brood.

lesser sickles \ *See: sickles*

lethal gene \ A gene that can cause a chicken to die, typically as an embryo during incubation. More than 50 different lethal genes have been identified in chickens, most of which are recessive. When two chickens are mated that carry the same lethal recessive, 25 percent of their offspring will display the lethal trait. They are readily recognized in any embryos that survive until hatch because they are usually associated with such oddities as stickiness, winglessness, twisted legs, missing beaks, or extra toes. A well-known lethal trait is the so-called creeper gene carried by short-legged Japanese chickens, once prized as broodies because their short legs keep their bodies close to the ground. Other known lethals are carried by Araucana (associated with ear tufts), dark Cornish (short legs), black Minorca (short legs, extra toes), and white Wyandotte (recessive white gene associated with early embryo death). Fortunately, lethal genes are relatively rare.

life cycle, direct \ A life cycle in which a parasite living in the body of a chicken lays eggs that are expelled in

LICE

These small, wingless, parasitic insects live on the skin of chickens. Lice chew on a chicken, causing the bird to break off or pull out its feathers trying to stop the irritation. The resulting damage makes the plumage look dull or rough. Louse-infested chickens don't lay well and have reduced fertility.

Lice spread through contact with an infested bird or its feathers and live their entire lives on a bird's body. You can easily see the straw-colored pests scurrying around on a chicken's skin, the scabby, dirty areas they create around the vent and tail, and louse eggs (called nits) clumped in masses around the feather shafts. A chicken with a properly shaped beak can minimize lice on its body through grooming; a chicken that has been debeaked or has an overgrown beak is more likely to have lice because it cannot groom properly.

An effective way to reduce the louse population is to diligently rake up and remove potentially nit-laden feathers from the house and yard. Treatment involves applying a delousing product approved for poultry to the shelter walls, roosts, nest boxes, floors, and chickens. Since this treatment won't kill nits, it must be repeated two more times at seven-day intervals to kill lice that hatch between times.

the chicken's droppings, and when those eggs are eaten by the same or a different chicken, the parasite eggs hatch and infect (or reinfect) that chicken.

life cycle, indirect \ A life cycle in which a parasite egg expelled in a chicken's droppings must be eaten by some other creature, such as an ant, a grasshopper, or an earthworm before it can infect (or reinfect) a chicken that eats the creature containing the parasite egg. *See also: intermediate host*

life span \ *See: longevity*

light \ *See: controlled lighting*

light breed/lightweight breed \ A breed that tends to be more streamlined than roundish and in which the hens generally weigh 4.5 pounds (2 kg) or less and the cocks generally weigh 6 pounds (2.7 kg) or less. Many of these breeds were developed primarily for egg production, although some are largely ornamental. Most light breeds lay eggs with white or creamy-white shells and tend to be somewhat flighty.

light meat/dark meat \ *See: dark meat/light meat*

limberneck \ *See: botulism*

lime sulfur \ A product sold as a veterinary dip and also to control bacteria and fungi on fruit trees. A 5 percent solution of lime sulfur, applied once weekly for four weeks, is an old-time remedy for treating mites.

line \ *See: strain*

line breeding \ A form of pedigree breeding, in which the influence of a superior sire or dam is concentrated by mating the bird to his or her best descendants. Pullets are mated to their sires or grandsires, cockerels are mated to their dams or grandams.

A good line-breeding plan includes four or more related families, starting with the best cock and four best hens available. Each family line consists of all the female offspring from its foundation hen. If one line fails to live up to expectations, it may be discontinued and replaced with a new line started with a foundation female from another line within the same strain.

In most cases both desirable and undesirable genes begin concentrating within three years. The third year is therefore when novice breeders, especially those who maintain too few lines, tend to get discouraged. The more separate lines are maintained, the greater the chance of producing at least one successful line. Maintaining several lines also helps preserve genetic diversity.

linseed \ *See: flaxseed*

linseed oil \ Oil extracted from linseed, which may be applied to cleaned roosts, nests, and cracks in walls or floors as a messy but effective way to rid housing of parasites that spend part of their time off a bird's body. Any other natural oil works equally well, but take care when using oil in a wood building, as it can create a fire hazard.

litter \ *See: bedding*

longcrower \ One of several breeds selectively bred for the sound or duration of their crow. These breeds generally have an upright stance, long legs, and long necks. They are known in many countries but likely all evolved from Japanese longcrowers, which in turn have their origins in the Shamo breed.

Japan recognizes three major longcrower breeds: The *Tomaru* (black crower) is noted for its rich two-tone call that deepens toward the end. In pitch it is intermediate between the calls of the other two breeds. The *Koeyoshi* (good crower), supposedly developed by crossing the *Tomaru* with the Plymouth Rock, has a deeper voice. The *Totenko* (red crower) is noted for its long tail, as well as for the duration of its high-pitched crow.

The call of a Japanese longcrower lasts at least 15 seconds; some go on for a full minute. It starts out sounding like the common cock's crow, but the final note is sustained (like a drawn-out train whistle) before petering out as the cock appears to run out of breath. The three distinct parts of the crow are called *dashi* (the beginning), *hari* (the stretch), and *hiki* (the finish).

The duration of crowing is preserved through generations by constant selection for the longest-crowing males. Unfortunately, the better the crowing ability of the breeding stock, the lower the fertility of the eggs and the more readily the delicate chicks succumb to various diseases. So although longcrower breeds appear in many countries, they remain relatively rare.

lopped comb

longevity \ The lifespan of a chicken under normal circumstances, which is about 12 years, although the occasional well-cared-for chicken may survive as long as 25 years. Most chickens don't live out their natural lives because they are eaten by a human or a predator or fall victim to disease, fatal injury, or poisoning. Although hens lay fewer eggs as they age, they become valuable breeders as long as they do lay, since their longevity demonstrates a hardiness that their offspring are likely to inherit.

longtail \ One of several breeds selectively bred for growing particularly long tail feathers. The cocks of these breeds have a nonmolting gene that causes some of their tail feathers to not molt, as well as a gene for rapid tail-feather growth. In Japan, where longtails have been developed to a fine art, tail feathers grow more than 3 feet (0.9 m) a year: a 10-year-old cock may thus have a tail longer than 30 feet (9 m). To attain this length each cock is housed separately in a tall cage with a high perch, and its tail feathers are protectively wrapped.

Longtails in North America include Cubalaya, Phoenix, Sumatra, and Yokohama. These breeds do not grow the luxuriant tails of Japan's Onagadori, the tail feathers of which are a minimum of 6½ feet (2 m) long, since they lack some of the genetic factors controlling the growth of excessively long tails, including full expression of the nonmolting gene that results in the shedding of tail feathers every year or two. Also, North American chicken keepers rarely spend enough time pampering cocks to keep their long feathers from breaking.

L

loose-feathered/loosely feathered \ Having feathers that are not held close to the body, creating something of a fluffy appearance that generally makes the chicken look bigger than it actually is. Cochins, Faverolles, and Orpingtons, for instance, are loosely feathered compared to tight-feathered breeds such as Buckeyes, Dorkings, and New Hampshires.

loose housing \ A building in which chickens are confined where they are free to roam within the walls. Loose housing is most commonly used for raising broilers or breeders but may also be used to maintain any flock during cold or wet weather.

lopped comb \ A comb that falls over to one side, as is typical of New Hampshire hens and hens of the Mediterranean breeds but is undesirable in cocks of those same breeds. A lopped comb is undesirable in both hens and cocks of any other single-comb breed, as well as in any breed with a pea or rose comb.

lower thigh \ The part of a chicken's leg between the shank and the knee joint. *[Also called: drumstick]*

luster \ The bright, glossy appearance of the plumage of a chicken, especially a cock, in outstanding condition. Luster results from proper nutrition, good health, and careful grooming.

lutein \ *See: xanthophyll*

lytes \ *See: electrolytes*

longtail rooster, in the style of Japanese ink wash paintings

L

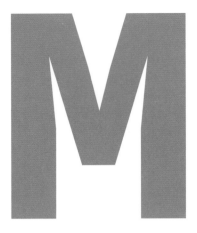

machine picking \ Using a picking machine to remove the feathers of recently killed chickens being butchered for meat. Machine picking can reduce plucking time to as little as half a minute, but it toughens the meat. Hand-picking is therefore more suitable for older chickens that tend to be tough already.

mail-order chicks \ Chicks shipped by mail within hours after they hatch. During their first few days of life, chicks obtain nourishment by continuing to absorb residual yolk, which in nature allows early-hatching chicks to remain in the nest under a hen until all her eggs hatch.

Although newly hatched chicks are resilient and ordinarily travel well, stress can be minimized by picking them up at the post office and getting them home quickly, rather than letting them bounce around in the mail carrier's vehicle until they are delivered. At the post office open the box in front of a postal employee in case you need to verify a claim for losses. As soon as possible at home, help the chicks rehydrate by dipping each chick's beak into water, warmed to room temperature, before putting it into a warm brooder.

main sickle \ *See: sickles*

main tail feathers \ The straight, stiff feathers of the tail. On a hen, the main tail feathers are below and between the coverts. On a cock, they are below and between the coverts and sickles. The main tail feathers are aerodynamic and when spread help a chicken fly.

mail-order chicks

main tail feathers

Malay \ An ancient breed from Southeast Asia, named after the Malay Peninsula. The Malay is the tallest breed — thanks to its long neck and long legs, combined with an upright stance — standing as tall as 2½ feet (0.75 m). It is considered a game breed and has the mean expression of a fighter, but it is raised primarily for its lean meat or for exhibition. It is a large, heavy chicken with a strawberry comb, beetle brow, short beak, downward-turning tail, and hard feathers. The Malay may be large or bantam and comes in a limited number of color varieties. The hens are poor layers of eggs with tinted shells and, when they are inclined to brood, make good mothers.

manure \ See box on page 172.

manure balls \ Pellets of dried droppings that cling to the toes of a chick (and occasionally a mature chicken). Manure balls develop on chicks that are overcrowded or kept in a brooder that is not regularly cleaned, but they may also result from crooked toes that force a chick to walk on the sides of its feet. Manure balls invite toe picking and can result in crippling. To remove them, set the chick's feet in warm, not hot, water just long enough for the hardened manure to soften, then gently pry the balls away from the toes.

manure box \ Droppings pit.

Malay hen

Malay rooster

M

Manx Rumpy
hen

Manx Rumpy \ An ancient breed from Persia (present-day Iran, Iraq, and surrounding areas) not commonly found in North America. The name comes from the similarity between its downturned rump and that of a Manx cat. It's a small chicken that looks like an Old English Game without a tail, and it comes in a similar range of colors. The hens are good layers of brown-shell eggs and will brood, but fertility is poor, which likely accounts for the scarcity of this breed. *[Also called: Manx Rumpie; Persian Rumpless; Rumpless Game]*

MANURE

Droppings, usually mixed with bedding, can be used to fertilize soil. The fertilizer value of chicken manure varies with its age and with the nutritional intake of birds at various stages of growth, as indicated below.

FERTILIZER VALUE OF CHICKEN MANURE

Chicken manure, by degree of freshness	% Nitrogen (N)	% Phosphate (P)	% Potash (K)
WET, STICKY, CAKED	1.5	1.0	0.5
MOIST, CRUMBLY TO STICKY	2.0	2.0	1.0
CRUMBLY BUT NOT DUSTY	3.0	2.8	1.5
DRY, DUSTY	5.0	3.5	1.8
Fresh manure, by age of birds			
BABY CHICK	1.7	1.3	0.7
GROWING CHICK	1.6	0.9	0.6
HEN	1.1	0.8	0.5

From: *Storey's Guide to Raising Chickens,* 3rd edition

M

Marans \ A dual-purpose breed developed in France and named after the port town of Marans. It comes in several color varieties, may be large or bantam, and has a single comb and a short tail held high. The original French strain has lightly feathered legs, while the British strain is clean legged.

The hens are good layers that produce eggs with dark chocolate-brown shells — the darkest shell of any breed — and some individuals lay eggs with speckled shells. Hens may brood, although many breeders discourage broodiness because it interferes with production of the unusually dark-shelled eggs, which generally bring a premium price.

Marek's disease \ A viral infection that primarily affects the nerves of growing chickens, causing leg paralysis and droopy wings, and sometimes death. Chickens with this disease shed the virus whether or not they show signs and thereby contaminate the yard for all future chickens.

Vaccination does not prevent chickens from becoming infected and shedding the virus, but it does prevent paralysis. For the vaccine to be effective, chicks must be vaccinated before being exposed to the virus, and therefore most hatcheries offer to vaccinate chicks before shipping them.

mash \ A mixture of feedstuffs that have been ground to various degrees of coarseness but are left recognizable enough that chickens can pick out what they like. Mash is most commonly available at a mill that does not have equipment to extrude pellets and is also the typical form of home-mixed rations.

Master Breeder \ An award conferred by the American Bantam Association to members who win 20 championships in a single variety of one breed in classes of one hundred birds or more, over a period of not less than five years.

Master Exhibitor \ An award conferred by the American Poultry Association to members who accumulate one hundred points by winning a sufficient number of class championships, the point values of which vary with the intensity of the competition. \ An award conferred by the American Bantam Association to members who win 20 championships in classes of one hundred birds or more.

Marans hen

egg with chocolate-brown shell

mate \ To pair a rooster with one or more hens for the purpose of producing fertile eggs for incubation. \ A hen or rooster so paired.

mating behavior \ Instinctive patterns of action displayed by chickens prior to copulation. A cock intent on mating an unfamiliar hen will likely chase after her, while the hen's response is to try to get away. The classic courtship behavior of a cock is to flick one wing along the ground while dancing around the hen, and she may or may not ignore his advances.

A dominant cock is likely to come up behind a hen with his head up and hackles out, whereupon the hen will crouch for mating. The cock stands on the hen's back, grabs her with his beak (by her comb, neck feathers, or skin), treads, then dips his tail to one side of the hen's tail while spreading his tail feathers so their cloacae can come together for the transfer of sperm. After mating, the rooster lets go, and the hen stands up and shakes her feathers back into place.

mating frequency \ The frequency at which naturally breeding chickens mate. The greatest frequency of mating is in early morning and late afternoon. Depending on the mating ratio and the number of other roosters providing competition, a cock may mate 30 or more times each day. Treading and neck grabbing may result in defeathering, and even injury, to hens if the mating ratio is too low (too many cocks for the existing hens), causing each hen to be mated too often, or if a cock repeatedly mates the same few hens to the exclusion of others.

mating ratio \ The number of hens available per cock. On average the optimal ratio for heavier breeds is 8 hens per cock, although a cock in peak form can handle up to 12. The optimal ratio for lightweight laying breeds is up to 12 hens per cock, yet an agile cock may accommodate 15 to 20. The mating ratio for bantams is 18 hens per cock, although an active cockerel might handle as many as 25. An older cock or an immature cockerel can manage only half the hens of a virile yearling.

If too many cocks are present, fertility will be low, because the cocks will

MATING RATIOS

For One Cock of the Breed	Optimal Number of Hens	Maximum Number of Hens
BANTAM	18	25
LIGHT BREEDS	12	20
HEAVY BREEDS	8	12

From: *Storey's Guide to Raising Chickens,* 3rd edition

M

spend too much time fighting among themselves. If cocks are too few, fertility will be low because the cocks can't get around to all the hens. A single cock with too many hens will favor some hens and ignore others. A cock with too few hens may cause injuries from treading. Where fertilized eggs are not needed for hatching, or the zoning ordinance does not allow roosters, hens can get along fine with no cock at all.

mating saddle \ *See: saddle*

maturity \ The stage at which a chicken reaches full size and stops growing, as distinct from sexual maturity. Large breeds mature, on average, in 8 to 10 months, bantams in 6 to 7 months.

mealiness \ An undesirable dusting of lighter-colored specks on the plumage of a solid buff or red variety. Mealiness differs from stippling in being less evenly spread and more spotty.
See also: stippling

meat bird \ *See: broiler*

meat breed \ A breed developed for rapid growth and heavy muscling. Although any healthy chicken may be raised for meat, some breeds are more suited to efficient meat production than others. These breeds share the following characteristics:
- They grow and feather rapidly
- They are broad breasted
- They are deep bodied (distance between back and keel)
- Their heart girth is spacious (circumference just behind the wings)

- Their backs are wide and flat (the same width front to back, not tapered)

Breeds originally developed for meat include Brahma, Cochin, and Cornish. American utility breeds with the greatest potential for efficient meat production are the Delaware, New Hampshire, and Plymouth Rock. Although the Jersey Giant grows to be the largest of all chickens, it is not economical as a meat breed because it puts growth into bones before fleshing out, taking six months or more to yield a significant amount of meat for its size.

Hybrids developed for meat production convert feed to muscle more efficiently than straightbreds. Fast-growing industrial Cornish Rock hybrids are the most efficient for rapid growth. These strains have white feathers for clean picking, in contrast to the hybrid broilers raised according to the Label Rouge model, which generally do not have white plumage. Pastured hybrids grow more quickly than straightbreds, making them more efficient for meat production, but grow more slowly than industrial broilers, making their meat more flavorful.

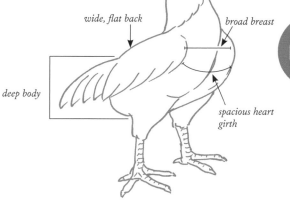

meat breed characteristics

wide, flat back

broad breast

deep body

spacious heart girth

M

BREEDS TYPICALLY RAISED FOR MEAT

Average Age at Slaughter	Breed	Average Weight at Slaughter (cock/hen)	Average Weight at Maturity (cock/hen)
12–16 weeks	DELAWARE	7½ lb/5½ lb	8½ lb/6½ lb
	NEW HAMPSHIRE	7½ lb/5½ lb	8½ lb/6½ lb
	PLYMOUTH ROCK	8 lb/6 lb	9½ lb/7½ lb
16–20 weeks	AMERAUCANA	5½ lb/4½ lb	6½ lb/5½ lb
	AUSTRALORP	7½ lb/5½ lb	8½ lb/6½ lb
	BUCKEYE	8 lb/5½ lb	9 lb/6½ lb
	CHANTECLER	7½ lb/5½ lb	8½ lb/6½ lb
	DOMINIQUE	6 lb/4 lb	7 lb/5 lb
	FAVEROLLE	7 lb/5½ lb	8 lb/6½ lb
	HOLLAND	7½ lb/5½ lb	8½ lb/6½ lb
	MARANS	7½lb/6 lb	8½ lb/7 lb
	RHODE ISLAND RED	7½ lb/5½ lb	8½ lb/6½ lb
	RHODE ISLAND WHITE	7½ lb/5½ lb	8½ lb/6½ lb
	SUSSEX	7½ lb/6 lb	9 lb/7 lb
	WYANDOTTE	7½ lb/5½ lb	8½ lb/6½ lb
20–24 weeks	BARNEVELDER	6 lb/5 lb	7 lb/6 lb
	CREVECOEUR	7 lb/5½ lb	8 lb/6½ lb
	DORKING	8 lb/6 lb	9 lb/7 lb
	HOUDAN	7 lb/5½lb	8½ lb/6 lb
	JAVA	8 lb/6½ lb	9½ lb/7½ lb
	NAKED NECK	7½ lb/5½ lb	8½ lb/6½ lb
	ORPINGTON	8½ lb/7 lb	10 lb/8 lb
	WELSUMER	6 lb/5 lb	7 lb/6 lb
24 weeks (6 months) or more	BRAHMA	10 lb/7½ lb	12 lb/9½ lb
	COCHIN	9 lb/6½ lb	11 lb/8½ lb
	CORNISH	8½ lb/6½ lb	10½ lb/8 lb
	JERSEY GIANT	11 lb/8 lb	13 lb/10 lb
	LA FLECHE	7 lb/5½ lb	8 lb/6½ lb
	LANGSHAN	8 lb/6½ lb	9½ lb/7½ lb
	MALAY	7½ lb/5½ lb	9 lb/7 lb
	ORLOFF	6½ lb/5 lb	8 lb/6½ lb
	SHAMO	9 lb/5½ lb	11 lb/7 lb

M

meat class \ One of several categories of chicken raised for meat, as established by the United States Department of Agriculture. All categories pertain specifically to Cornish Rock hybrids, the progeny of a cross between a purebred Cornish cock and a purebred Plymouth Rock hen, both from specialized strains. Despite the standard practice of designating the cock's breed first, the USDA lists the hen's breed first.

ROCK CORNISH GAME HEN, CORNISH GAME HEN. Not a game bird at all and not necessarily a hen, but an immature (usually five to six weeks old) Cornish or Cornish-cross chicken of no more than 2 pounds (0.9 kg) dressed weight. This single-serving chicken is typically stuffed and roasted whole.

BROILER, FRYER. A tender chicken, usually less than 13 weeks of age, that has soft, pliable, smooth-textured skin and a flexible breastbone. A broiler/fryer is tender enough to be cooked by any method.

ROASTER, ROASTING CHICKEN. A tender chicken, usually between three and five months of age, that has soft, pliable, smooth-textured skin and a breastbone that is somewhat less flexible than that of a broiler. A roaster is usually stuffed, roasted whole, and sliced for serving.

CAPON. A castrated cockerel, usually less than eight months of age, with tender meat and soft, pliable, smooth-textured skin. A capon weighs more than a roaster and is prepared in the same way.

HEN, FOWL, BAKING CHICKEN, STEWING CHICKEN. A mature hen, usually older than 10 months of age, with meat that is less tender than that of a roaster and a nonflexible breastbone. This chicken must be cooked by a moist method such as stewing, braising, or pressure cooking. Stewing hens are generally laying hens that are no longer economically productive.

COCK, ROOSTER. A mature male chicken with coarse skin, a hardened breastbone tip, and tough, dark meat. Such a chicken is generally not fit to eat, although with long, moist cooking may be made tender enough to chew.

meat grading \ An indication of the level of quality of chicken meat, according to definitions established by the United States Department of Agriculture. Official grading is a voluntary service paid for by poultry processors as an aid in marketing their products. Chicken meat sold by these companies has a grade stamp within a shield on the package wrapper, or sometimes on a tag attached to a wing.

Grade A is the highest quality, indicating the chicken is free of major defects, Grade B indicates moderate defects, and Grade C indicates abnormal conformation and serious defects. Grade A is sold at the meat counter. If an entire chicken doesn't qualify as Grade A, it is cut up and the qualifying parts sold as Grade A. Otherwise it is processed into canned chicken, soup, and similar products.

meat spot \ A blemish that appears within an egg as a brown, reddish brown, tan, gray, or white spot, usually on or near the yolk. A meat spot may be a tiny piece of reproductive tissue that tore loose as the egg was forming or may have

M

started out as a blood spot that changed color because of a chemical reaction. Meat spots look unappetizing but do not make eggs unsafe to eat. They are relatively uncommon and are likely hereditary — if you hatch eggs with meat spots, the resulting hens will probably lay eggs containing meat spots.

mechanical transmission \

The spreading of a disease by means of pathogens carried on the surface of something (called a vector) that might be either animate or inanimate. Animate vectors include the feet of flies, rodents, or wild birds. Inanimate vectors include dirty mud puddles, contaminated injection needles, and any kind of fomite.

medicated feed/ration/ starter \ Commercially prepared

chick feed containing a coccidiostat, the only drug added to packaged chicken feed sold for backyard use. Chicks should not need medicated starter if:

- They hatch in late winter or early spring (before warm weather allows coccidia to thrive)
- They are not crowded
- They are kept on dry, clean litter
- They always have fresh, clean drinking water

They definitely should not be fed medicated ration if they have been vaccinated with an anticoccidial vaccine, since the coccidiostat in the starter would neutralize the vaccine. Using medicated starter is a good idea if:

- Chicks are brooded in warm, humid weather
- A large number of chicks are raised together

- They are brooded in the same place for more than three weeks
- The brooder is used for one batch of chicks after another
- Brooding sanitation is less than optimal
- You are raising chicks for the first time
- You are not raising chickens for meat (otherwise, you'll need an alternative feed for the withdrawal period)

Mediterranean class \ One of

six groupings into which the American Poultry Association organizes large chicken breeds. The breeds in this class originated primarily in Italy and Spain. The breeds from Italy are Ancona, Leghorn, and Sicilian Buttercup; the breeds from Spain are Andalusian, Catalana, Minorca, and Spanish. Mediterranean breeds lay white-shell eggs and tend to be particularly prolific but are also high strung.

mice \ *See: rodents*

microflora \ *See: intestinal microflora*

milk \ To collect semen for use in artificial insemination. \ The semen thus collected. \ A dairy product chickens love and that is good for them. Milk is 87 percent water, and the remainder is loaded with protein, carbohydrate, fat, vitamins, and minerals. Milk helps put weight on meat birds, but layers that drink too much get fat, and fat hens don't lay well. A good rule of thumb is to feed hens no more than 10 pounds (4.5 kg) of liquid milk per 50 pounds (22.7 kg) of rations they consume.

mille fleur

mille fleur \ A color pattern, from the French term *mille fleurs*, meaning thousand flowers. Each mahogany feather is tipped with a crescent-shaped black bar and a V-shaped white spangle. Belgian Bearded d'Uccles and Booted Bantams are commonly called bearded and nonbearded mille fleurs, respectively, because mille fleur is the most popular color variety for both breeds, even though both come in other color varieties and the mille fleur color pattern occurs in other breeds.

millie \ Short for mille fleur.

mineral grit \ *See: calcium grit*

miniature \ A breed of bantam that has a large counterpart, as distinct from a true bantam, which does not.

Minorca \ An old breed originating in Spain and named after the Spanish island of Minorca. It is the largest Mediterranean breed, comes in both rose-comb and single-comb varieties and a few color varieties — of which black is the original and still the most common — and may be large or bantam in size. The Minorca has a large white earlobe and looks so much like the white-faced black Spanish that it is sometimes referred to as the red-faced black Spanish.

The Minorca does well in hot climates but is not cold hardy because its large comb and wattles freeze easily. The hen's comb is somewhat unusual in that it folds back on itself, rather than drooping over to one side, as is common among laying breeds. The hens are good layers, producing eggs that are among the largest of those with white shells, and rarely brood.

Minorca hen

M

mite \ A tiny spiderlike body parasite of the class Arachnid, related to the tick, that has four pairs of legs as an adult. Several species of mite — including the red mite, northern fowl mite, and scaly leg mite — survive by consuming a chicken's skin, feathers, or blood. Mites cause irritation, feather damage, increased appetite, low egg production, reduced fertility, retarded growth, and sometimes death.

Modern Game \ A breed developed in England from the Old English Game for exhibition purposes. Despite its name, this breed was never intended for cockfighting. The Modern Game

Modern Game hen

is a tall, slender chicken with a long neck, long legs, compact body, thin pinched tail held nearly horizontally, and an upright stance, which combine to give this breed an elegant, statuesque appearance. It comes in a number of color varieties, may be large or bantam, and has a small single comb. To be exhibited, the cocks must have their combs, wattles, and earlobes dubbed at about eight months of age.

The Modern Game has short, hard, tight feathers and little fluff for insulation, making it vulnerable to cold weather. The hens are fair layers of eggs with tinted shells and are excellent broodies that sometimes must be discouraged from nesting repeatedly, as these small birds have few body reserves to call on.

Modern Game Class \ One of the exhibition categories designated by the American Bantam Association, consisting entirely of different varieties of the Modern Game breed.

molasses flush \ *See: laxative flush*

moldy feed \ *See: mycotoxin*

molt \ The periodic shedding and renewal of feathers. Short day lengths serve as a signal to wild birds that it's time to renew their plumage in preparation for migration and the coming cold weather. Like most birds, chickens molt at approximately one-year intervals, usually for 14 to 16 weeks during the late summer or early fall. Because molting occurs over a period of weeks, rarely does a chicken become completely naked.

M

During the molt all hens slow down in production, and some stop laying altogether. After the molt their feed efficiency improves, their eggs become larger, and egg quality is better than at the end of the previous laying period. On the other hand, they won't lay quite as well as they once did, and egg quality will decline faster than during the previous laying period.

The best layers molt late and fast. They lay eggs for a year or more before molting — which is why good layers look ragged — and take only two to three months to finish the molt. The poorest layers molt early and slowly. Some may lay for a few months, then go into a molt that lasts as long as six months — which is why the plumage of lazy layers nearly always looks shiny and sleek. Checking the wing molt is useful for determining whether a hen is a fast or slow molter.

Molting may occur out of season as a result of disease or stress, such as chilling or going without water or feed. A stress-induced molt is usually partial and does not always cause a drop in laying, while a normal full molt is typically accompanied by at least a slowdown.   *See also: wing molt*

molting sequence \ Molting

occurs in a specific sequence, starting at the head and neck and gradually working toward the tail. In a rapid molt some areas molt simultaneously.

mooney \ Description of spangles

that are undesirably round, instead of more properly V-shaped. Mooney spangles most commonly appear in golden and silver spangled Hamburgs.

morbidity \ The number of chickens

in a flock, scaled to the flock's size, that contract a disease.

morbidity rate \ The number of

chickens in a flock, scaled to the flock's size, that contract a disease within a given time period.

mortality \ The number of chickens

in a flock, scaled to the flock's size, that die from a disease.

mortality rate \ The number of

chickens in a flock, scaled to the flock's size, that die from a disease within a given time period.

molting sequence

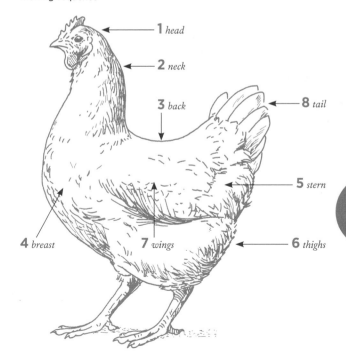

1 *head*

2 *neck*

3 *back*

8 *tail*

5 *stern*

4 *breast*

7 *wings*

6 *thighs*

M

mossy \ Irregular or blurred coloring that muddles the desired color pattern of the plumage, typically in laced varieties. On particolored plumage, mossiness consists of flecks of black color over white, buff, or gold feathers.

mottled \ An irregular pattern created by some of the feathers having white tips, as is typical of the plumage of Anconas, Houdans, and Javas. \ An undesirable color pattern of individual feathers that are spotted with colors or shades of color differing from the desirable ground color.

muff \ Feathers sticking out from both sides of the face below and around the eyes and covering the earlobes. A muff always occurs in association with a beard. In large breeds the beard and muff constitute three distinct clumps of feathers, while in bantams they run together into one continuous formation. Breeds with muffs include Ameraucana, Faverolle, and Houdan.
[Also called: whiskers]

mulberry \ See: gypsy

multiple spurs

multiple spurs \ A unique characteristic of Sumatras, in which mature cocks have three to five spurs vertically arranged on each shank and hens have three to five vestigial spurs in the form of flattened bumps.

mushy chick disease \ A condition in which a chick's yolk sac isn't completely absorbed so the navel can't heal properly; as a result bacteria invade through the navel, causing chicks to die at hatching time and for up to two weeks afterward. Mushy chick disease has a number of causes, including incubation of dirty eggs, operation of an unsanitary incubator, and improper temperature or humidity during incubation. [Also called: omphalitis]

mycotoxin \ Any of a number of poisons produced by molds that grow naturally in grains. Poisoning is difficult to identify and diagnose, in part because the feed may contain more than one kind of mycotoxin. A positive diagnosis usually requires analysis of the feed to identify any fungi present. All mycotoxins increase a chicken's need for vitamins, trace elements (especially selenium), and protein. Once the contaminated feed is removed, chickens usually recover.

To prevent mold from forming, store feed away from humid conditions and use plastic containers rather than metal ones, which generate moisture by sweating. Never feed chickens anything that is moldy.

Naked Neck hen

M

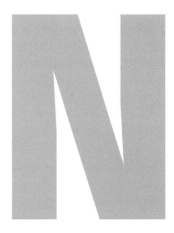

Naked Neck \ An old dual-purpose

breed originating in Transylvania (once part of Hungary, now part of Romania) and developed in Germany. The breed is sometimes called a Turken because it looks like a turkey crossed with a chicken. Naked Necks have a single comb, may be large or bantam, and come in a few color varieties. They are easy to identify by their red, featherless necks. In fact, they have less than half the number of feathers of breeds of comparable size, giving them three advantages:

- They have excellent feed conversion efficiency because they don't need as much dietary protein as breeds that grow more feathers
- They can handle hot weather better than more heavily feathered breeds, although they need shade to avoid sunburn to the unfeathered parts
- They are easy to pluck clean

The hens are good layers of brown-shell eggs, may brood, and make outstanding mothers.

[Also called: Transylvanian Naked Neck, Turken]

Nankin \ An ancient breed of true

bantam originating in England. Exactly how it got the name Nankin is not known. The name may derive from the French word *nain*, meaning dwarf, or from the British word *nankeen*, in reference to once-fashionable buff-color fabric imported from Nanking, China. The Nankin comes in a single color pattern — buff plumage set off by a full black tail — and may have either a single comb or a rose comb.

Nankins are somewhat noisy, are capable fliers, and are social in tending to forage together as a group. The hens lay round eggs with tinted shells, are good broodies, and are excellent mothers.

National Organic Program (NOP) \ The set of standards estab-

lished by the United States Department of Agriculture pertaining to organic certification. Since these standards leave many areas open to interpretation, the numerous state and private certifying agencies must interpret the standards and make judgment calls as to whether each poultry producer is or is not in compliance.

Nankin hen

N

National Poultry Improvement Plan \ A federal agency that works with state agencies to certify chicken flocks as being free of several serious diseases. Chickens purchased from a member are pretty certain to be healthy. On the other hand, a lot of poultry breeders don't want to get involved in government bureaucracy, which does not automatically mean their chickens are unhealthy.

natural \ A word often used to describe poultry meat and eggs that are not industrially produced but are not organically certified. Guidelines published by the United States Department of Agriculture state that natural products may undergo only minimal processing (such as being cut up into pieces and/or frozen) and cannot contain any artificial ingredients or preservatives. A label claiming a product is natural must briefly explain what is meant by the use of that word, but otherwise the claim that chicken meat or eggs are natural is unregulated — although that's subject to change.

NEST BOX/ NESTING BOX

A manufactured nest provides a place where hens will lay eggs that may easily be found, as opposed to a secluded place to lay that is chosen by a hen. Furnish one nest for every four to five hens in the flock.

A rail just below the entrance to the nest gives hens a place to land before entering. Place the rail far enough from the nest opening that any chicken roosting on it won't fill the nest with droppings. For most chickens the rail should be no closer than 8 inches (20 cm) from the edge of the nest.

NEST BOX SIZE GUIDELINES

Breed Size	Nest Box Size
Bantam	10" wide × 12" high × 10" deep (25 × 30 × 25 cm)
Light	12" wide × 14" high × 12" deep (30 × 35 × 30 cm)
Heavy	14" wide × 14" high × 12" deep (35 × 35 × 30 cm)

N

natural-draft incubator \ *See: gravity-flow incubator/gravity-ventilated incubator*

naturally raised \ A description of chickens raised without growth promotants, fed no animal by-products, and given no antibiotics with the exception of coccidiostats called ionophores, which are chemically similar to antibiotics. This definition, put forth by the United States Department of Agriculture, does not address such hot-button issues as confinement, genetic engineering, and the use of pesticides.

neck \ The part of a chicken connecting the head to the rest of its body, typically made up of the upper 14 vertebrae of the spinal column. This large number of vertebrae (compared, for instance, to a human's 7), along with the saddlelike shape of the vertebrae, give a chicken the great flexibility of movement that allows it to turn its head to look in all directions, preen all parts of its body (except the head and the back of the neck), peck for food, and engage in nest building.

neck wringing \ Killing a chicken by twisting, crushing, or stretching its neck. *[Also called: cervical dislocation]*

necropsy \ The examination of a dead chicken, or a recently killed diseased chicken, in an attempt to determine what was wrong with it (equivalent to a human autopsy).

necrotic enteritis \ Inflammation and decaying of intestinal tissue.

nematode \ A parasitic roundworm.

nest \ A secluded place where a hen feels she may safely leave her eggs. \ To brood.

nest egg \ A fake egg placed in a nest to encourage hens to lay their eggs where you can find them. Seeing what she believes is an egg already in the nest makes a hen feel it is a safe place to deposit her own egg. Imitation eggs, available at toy stores and hobby shops, should be made of wood or other heavy material; air-filled plastic eggs easily bounce out of a nest. Fake eggs need not be exactly the same size and color as the hens' own eggs. The size and weight of a golf ball makes it an ideal nest egg.

nesting call \ The sound made by a hen looking for a suitable place to lay eggs. Cocks make a similar, but more intensely excited, sound to show a hen a potential nest site. While gabbling, the cock nestles into the chosen spot as if he is about to lay an egg himself. Occasionally, a hen will check out the spot and respond with her own song. The sound is more common among pullets and cockerels but may also be made by a mature hen that resumes laying after a molt.

nesting material

nesting material \ Soft, clean bedding placed in a nest box to keep eggs clean and prevent breakage. Wood shavings, shredded paper, fine straw, or well-dried chemical-free lawn clippings all make good nesting material. A 4-inch (10 cm) sill along the bottom edge of the nest opening will hold in the nesting

N

material and keep eggs from rolling out. The nesting material must be changed as often as necessary, as dictated by the amount of hen activity, to keep eggs clean and unbroken.

nest pad \ A sheet of thick material of suitable size to fit the bottom of a nest box to keep it clean. Store-bought nest pads, such as those made of nonreusable excelsior (wood fiber) or cleanable plastic, are designed to replace nesting material. Homemade nest pads, such as might be cut from nonreusable corrugated cardboard or cleanable asphalt shingling, are used beneath nesting material to prevent bedding, dirtied with poop or broken eggs, from sticking to the nest bottom.

nest run \ Ungraded eggs.

New Hampshire hen

Newcastle disease (ND) \ A contagious viral infection that affects a chicken's respiratory system and sometimes the nervous system. Three types of virus cause Newcastle with varying degrees of severity. The mild form is easily confused with other respiratory diseases. Most chickens survive and thereafter become immune, but remain carriers for a few weeks after recovery. The severest form, called exotic (because it originated outside the United States) or velogenic Newcastle, is one of the most devastating diseases of poultry worldwide and can result in a death rate as high as 100 percent. A vaccine is available for use in areas where Newcastle is common.

New Hampshire \ A dual-purpose breed developed in the United States and named after its state of origin. New Hampshires were created through selective breeding of Rhode Island Reds to improve the rate of growth and degree of muscling for use in the broiler industry and are therefore blockier than Rhode Island Reds. They have a single comb, may be large or bantam, and come in one color variety — a light reddish bay that's more golden than the Rhode Island's rich mahogany plumage. The hens are decent layers of eggs with brown shells, may brood, and when they do they make good mothers.

New Hampshire Red \ Incorrect designation for New Hampshire.

nictitating membrane \ *See: eyelid*

nidifugous \ Capable of leaving the nest soon after birth, from the Latin words *nidus*, meaning nest, and *fugere*, meaning flee.   *See also: precocial*

nits \ The eggs of lice.

noninfectious disease \ An illness that is not caused by a biological organism but by some nutritional deficiency or excess, chemical poisoning, traumatic injury, or excessive stress.

northern fowl mite \ A dark red or black mite that is active in cool climates during winter and may be seen crawling on eggs in nests or on birds during the day, causing scabby skin and darkened feathers around the vent. Since these mites increase rapidly, act fast by dusting birds and nests with an approved insecticide. Northern fowl mites cannot live long in an unoccupied chicken house, which will be free of them after about three weeks.
[Also called: feather mite]

Norwegian Jaerhon \ A layer breed developed in Norway's Jaer district from a landrace breed that existed prior to the importation of chickens from outside the country. The Jaerhon is a small, hardy chicken with a single comb and may be large or bantam. It comes in two recognized varieties, light (or light yellow) and dark (or dark brown), but also typically throws sports from which other color varieties are being developed. The breed is autosexing: Light chicks are yellow, and the pullets have a brown stripe running down the back; dark chicks are

Norwegian Jaerhon rooster

N

brown with a yellow head spot, the cockerels' spot being larger than the pullets'.

The hens are good layers of white-shell eggs and seldom brood. They are capable fliers and require a high fence or a covered run to keep them from flying outside the yard to lay.

nostrils \ Two openings at the base of the upper half of a chicken's beak that serve the purpose of admitting air into the respiratory system. Most chickens have slit-shaped nostrils. Duplex combed chickens (other than Buttercups) are the exception: they have cavernous nostrils.

not show quality (NSQ) \ A chicken that does not conform closely enough to the standard description for its breed and variety to compete favorably in an exhibition. Birds that are not show quality may be either breeder quality or pet quality; otherwise they are culls.

notifiable disease \ *See: reportable disease*

no-yolker \ *See: yolkless egg*

nostrils

Old English Game hen

Old English Game rooster

N

obesity \ *See: fat pad*

offal \ Inedible organs, such as the crop and intestines.

off feed \ Having little or no appetite, therefore failing to obtain adequate nutrition. Anything that causes hens to eat less than usual also causes them to lay less than usual. Likewise, anything that causes broilers to eat less than usual reduces their rate of growth. To encourage eating, provide feed more frequently or stir the ration between feedings.

offset wire \ An electrified wire running alongside a nonelectric fence and held several inches away from it by extra-long insulators, or offset brackets. On a chicken fence, an offset wire is used primarily as a scare wire to discourage predators.

oil gland \ *See: uropygial gland*

Old English and American Game Class \ One of the categories into which the American Bantam

Association organizes bantams. This class includes all varieties of Old English and American Game, but not Modern Game, which has its own separate class.

Old English Game (OEG) \

An ancient breed from England, originally kept primarily for cockfighting but now considered ornamental. The Old English Game is a smallish chicken with a single comb, comes in a sizable selection of color varieties, and may be large or bantam. Similar to the red jungle fowl (and likely a direct descendant), it is the active, hardy, feral breed most commonly seen foraging along North America's rural roads and roosting in trees. To be exhibited the cocks must have their combs, wattles, and earlobes dubbed. The hens are decent layers of eggs with tinted shells and make good broodies.

omega-3 fatty acids \

Polyunsaturated fatty acids found in a higher rate in the yolks of eggs laid by pastured hens than in those laid by confined hens. Omega-3s are required for human growth, development, and good vision; deficiency has been linked

offset wire

to heart disease, cancer, Alzheimer's disease, and other devastating illnesses. Although the yolks' total fat content remains the same as if the hens were not on pasture, the percentage of polyunsaturated fat increases. As an alternative to pasture, omega-3 may be boosted in eggs by adjusting the hens' diet to include 10 percent flaxseed — but no more, or eggs may take on a fishy flavor.

omphalitis \ *See: mushy chick disease*

Onagadori \ *See: longtail*

oocyst \ An infective fertilized egg of certain one-celled parasites, including the protozoa that cause coccidiosis in chickens. A typical oocyst contains eight coccidia, which are released when the oocyst is eaten by a chicken and gets crushed by the chicken's gizzard.

open show \ A show at which both adults and youths compete against each other. \ A show at which coop tags are left open during judging.

operant conditioning \ A training technique, successfully used with chickens, whereby a bird is offered a reward (or positive reinforcer) when it performs a desired behavior. A positive reinforcer may be anything a chicken wants, seeks, or needs — most commonly, some tasty bit of food. The idea is that a chicken is more likely to repeat a desired behavior when it associates its action with the pleasure of obtaining a reward, whereas a chicken that does something other than the desired behavior, and obtains no reward, is less

likely to repeat that action. Eventually, the chicken learns what, exactly, it is supposed to do by avoiding behavior that results in no reward.

opportunistic \ Exploiting circumstances to gain immediate advantage. Chickens are opportunistic feeders, inasmuch as they eat whatever is available, whether derived from plants, insects, or other animals. A disease or infection is opportunistic when it is caused by an organism that is normally harmless but becomes activated when a chicken's immune system is otherwise impaired.

organic \ Determining what exactly "organic" means has been problematic since the word came into common use. People who raise chickens for their own purposes can set their own standards, but those who sell poultry meat or eggs labeled as being organic must by law be certified.

organic certification \ Verification by an outside agency as having been produced according to organic specifications. Numerous state and private certifying agencies accredited by the United States Department of Agriculture all have their own set of standards, developed as their individual interpretations of the National Organic Program and therefore differing slightly from one agency to another. All certifiers, however, agree that organic chickens must be raised in uncrowded, humane conditions and fed a vegetarian, antibiotic-free diet. To be formally certified as organic requires signing up with an accredited certifying agency, keeping detailed records, submitting an annual farm plan, and being inspected annually to verify that organic practices are being followed.

Orloff \ Ancient breed originating in Persia (now Iran), developed in Russia, and named after the Russian who popularized the breed, Count Orloff-Techesmensky. This large, tall, dual-purpose breed has a small walnut comb, a beetle brow, and a full beard and muffs, and seems to have an undersized head with what some poultry enthusiasts consider a gloomy expression. The breed is extremely cold hardy, comes in a few color varieties, and may be large or bantam. The hens are good layers of eggs with tinted shells and may brood.

ornamental breed \ Any breed that stands out for aesthetic qualities, rather than for its efficient production of eggs or meat. Because of their size, and the small size of their eggs, bantams are considered to be ornamental no matter how well they lay or how fast they grow. **See box on page 192 for characteristics of specific ornamental breeds.**

Orpington \ A dual-purpose breed developed in England and named after the town of Orpington in Kent. Orpingtons have a single comb, may be large or bantam, and come in a few solid color varieties, of which buff is the most common. This hefty breed is loosely feathered, making it cold hardy and giving it the appearance of being more massive than it actually is. The hens are good layers of eggs with brown shells, tend to go broody, and make good mothers.

outcross \ To breed chickens that are not directly related. By increasing heterozygosity, outcrossing results in hybrid vigor but may also introduce undesired genetic traits. \ The offspring of chickens that are not directly related.

Orloff hen

Orpington hen

ORNAMENTAL BREED CHARACTERISTICS

AMERICAN GAME BANTAM. Small size; long hackles, saddles, and sickles

BELGIAN BEARDED D'ANVERS. Beard; thick muffs; forward stance

BELGIAN BEARDED D'UCCLE. Beard; boots; vulture hocks; squat stance

BOOTED BANTAM. Boots; vulture hocks; squat stance

COCHIN. Ball shape resulting from long, dense plumage

CREVECOEUR. Crest; jet-black plumage

CUBALAYA. Long hackles and saddles; upright stance

DUTCH BANTAM. Tiny size; large comb; long tail; upright stance

HOUDAN. Crest; V-comb; five toes

JAPANESE BANTAM. Short legs; large tail; large comb; forward stance

KRAIENKOPPE. Beetle brow; small comb and wattles; upright stance

LA FLECHE. Hornlike V-comb; white earlobes; jet-black plumage

LANGSHAN. Leg feathering; full, high tail; tall, upright stance

MODERN GAME. Long neck; long legs; horizontal tail; upright stance

NAKED NECK. No neck feathers; minimal plumage

NANKIN. Small size; golden chestnut feathers with black tail

OLD ENGLISH GAME. Small size; broad tail; diversity of plumage colors

ORLOFF. Tall standing; beetle brow; beard and muffs

PHOENIX. Small size; cock's long tail

POLISH. Large crest; V-comb; sometimes beard and muffs

REDCAP. Outsize rose comb

ROSECOMB. Large rose comb; large white earlobes; broad tail

SEBRIGHT. Small size; laced plumage; hen feathered

SERAMA. Smallest breed; enormous range of plumage colors

SICILIAN BUTTERCUP. Cup-shaped, crownlike comb

SILKIE. Black skin; crest; fur-like feathers; five toes

SPANISH. Pure white face

SPITZHAUBEN. Forward-facing crest; V-comb

SULTAN. Crest; beard and muffs; feather footed; vulture hocks; five toes

SUMATRA. Black skin; lustrous plumage; flowing tail; multiple spurs

YOKOHAMA. Flowing tail; cock's long saddle feathers

outer thick \ The dense albumen that makes up the greater portion of the egg white. It is called the outer thick to distinguish it from the other thick layer of egg white, called the inner thick.

ova (singular: ovum) \
Undeveloped yolks, each consisting of a hen's reproductive cell that, upon being fertilized, is capable of developing into an embryo. The body of a newly hatched pullet contains all the ova that could possibly develop into fully formed eggs within her lifetime. Although each pullet starts with somewhere between two thousand and four thousand ova, rarely does a hen lay more than about a thousand eggs during her lifetime.

The ova are clustered, like a bunch of tiny grapes, approximately at the middle of a hen's backbone. Depending on the hen's age and how long she's been laying, her ova range from head-of-a-pin size to nearly the full size of a yolk in one of her eggs. In a pullet that hasn't yet reached the age of lay or a hen that's not laying (for instance, a molting hen or one of advanced age), all of the ova are tiny. When a pullet reaches age of lay or a hen resumes laying, her ova mature one by one, so at any given time her body contains ova at various stages of development. Approximately every 25 hours, one becomes mature enough to ovulate.

ovaphobia \ The rare condition of being irrationally afraid of eggs (from the Latin word *ova*, meaning eggs, and the Greek word *phobos*, meaning fear). Probably the least unusual fear of eggs is the result of media exaggerations concerning the cholesterol content of eggs and their potential for salmonella poisoning. The most famous person to truly suffer from ovaphobia was the horror-film maker Alfred Hitchcock, who once admitted to being afraid of eggs and said he particularly found the yolks to be revolting.

ovary \ A hen's reproductive organ in which ova are produced. Like humans and other vertebrates, a pullet starts life with two ovaries; as she matures, however, the right ovary remains undeveloped and only the left one becomes fully functional.

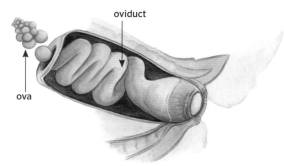

oviduct

ova

oviduct \ A tube starting at a hen's ovary, through which a yolk travels as it is developed into an egg ready to be laid. A hen's oviduct is similar to a woman's Fallopian tube.

oviposition \ To lay an egg. From the Latin words *ovum*, meaning egg, and *ponere*, meaning to place.

ovulate \ To discharge a mature ovum, or yolk, from the ovary into the funnel of the oviduct, a process that occurs in a laying hen approximately every 25 hours. Most hens ovulate within 1 hour of laying the previous egg.

pair \ A cock and hen of the same breed and variety.

parasite \ Any living thing that lives off another living thing (the host) without providing any benefit in return. Technically, all infections are caused by parasites of one sort or another. But more commonly a parasite is considered to be an animal form (such as a worm or a mite) that lives on or within another animal form (such as a chicken). Most chickens have some kind of animal-form parasites, which normally do not pose a serious threat. A heavy infestation of even the most benign parasite, however, causes stress that lowers a chicken's susceptibility to other infections.

parasite, external \ Any parasite that lives on or attacks the outside of a chicken's body, including mites; lice; and a host of fleas, flies, and other minor pests. Options available for dealing with external parasites include dust baths, lime sulfur, linseed oil, systemic inhibitors, and pesticides approved for use with poultry. When external parasites get out of control, sometimes a combination of methods is needed. **See chart on page 195 for diagnosing parasites.**

parasite, internal \ Any parasite that lives inside a chicken's body, usually in some part of the digestive tract. The most common internal parasites of chickens are worms and coccidia. Effectively controlling internal parasites requires knowing whether or not an intermediate host is involved. **See chart on page 195 for diagnosing parasites. See chart on page 153 for intermediate hosts.**

parasiticide \ A drug used to destroy parasites on or in chickens. Use only products approved for use with poultry or recommended by and used under the supervision of a veterinarian. The indiscriminate use of parasiticides can cause parasites to become resistant to future control.

parental immunity \ The resistance to disease passed from breeder cocks and hens to their offspring through the egg. Parental immunity is greatest in strong and vigorous breeder flocks, particularly in breeders that are at least two years old and still healthy, fertile, and laying well.

parson's nose \ *See: uropygium*

particolored \ Description of plumage having feathers of more than one color or more than one shade of a single color.
[Also called: pied]

SIGNS OF COMMON INTERNAL PARASITES

Parasite	Signs
WORMS	Pale head; droopiness; reduced laying; gradual weight loss; sometimes foamy diarrhea
COCCIDIA	Runny, off-color droppings, sometimes tinged with blood; slow growth; pale skin and shanks; reduced laying

From: *The Chicken Health Handbook*, 1994

SIGNS OF COMMON EXTERNAL PARASITES

Parasite	Signs
RED MITES	Red or black specks crawling on skin at night, hiding in nests and woodwork cracks during the day; pale comb and wattles, weight loss, death among young birds and setting hens
NORTHERN FOWL MITES	Red or black specks around vent or on eggs in nests, insect eggs along feather shafts; dirty-looking vent and tail, weight loss, drop in laying
SCALY LEG MITES	Swollen-looking shanks and toes with raised, crusty scales; mites are too small to see with the naked eye
LICE	Pale insects scurrying on skin, white eggs clumped at base of feathers; dirty-looking vent and tail area, weight loss, reduced laying and fertility

From: *Storey's Guide to Raising Chickens*, 3rd edition

PASTING

A fairly common condition in newly hatched chicks is having loose poop stuck to the vent area. Pasting may be caused by stress due to chilling or overheating (especially in mail-order chicks) or by feeding an improper ration. Soft droppings that stick to a chick's vent will harden and seal the vent shut, eventually causing death.

To remove the hardened droppings, run a little warm (not hot) water over the chick's bottom until the pasting has softened enough to be gently picked off without tearing the chick's skin. When the mess has been removed, gently dab the chick's bottom dry with a paper towel and apply a little Neosporin or Vaseline to protect the affected area and prevent a recurrence.

Pasting is less likely to occur in chicks that are drinking well before they start eating. Persistent pasting can sometimes be prevented by either feeding newly hatched chicks mashed hard-cooked egg and/or combining starter ration with finely crushed grains for their first few days of life; if that solves the problem, in the future switch to a different brand of starter ration that does not cause pasting. A probiotic or other supplement designed specifically for newly hatched chicks is also helpful.

[Also called: sticky bottom]

checking for pasting

P

pasture \ Land covered with grass and other short plants suitable for grazing. A mixed pasture is more nutritious than grass alone, and plants ordinarily considered weeds — including chicory, dandelion, and plantain — are enjoyed by and good for chickens.

Among warm-season greens, alfalfa is a good pasture choice where adequate rainfall or irrigation is available. Lespedeza has a similar nutritional value and grows well in southern regions, although in the colder north it must be seeded as an annual. Ladino and alsike clover are other popular warm-season choices.

Orchard grass is a cool-season pasture grass with a broad leaf that chickens like, and it gives them an early start on spring greens. A mixture of grasses extends the season and might include perennial ryegrass, fescue, Kentucky bluegrass, Canada bluegrass, and timothy. Any of the cereal grasses — such as barley, oats, rye, or wheat — make good cool-season pasture. Chickens plucking at the plants causes them to tiller, or grow more stems, so if the chickens are later removed and the cereals are allowed to go to seed, they'll produce more grain.

pasture confinement \ Keeping chickens enclosed within a portable floorless shelter that is periodically moved to fresh pasture. A 5-foot-by-6-foot (1.5 by 1.75 m) shelter can house up to 25 broilers or a dozen layers. During the first few moves, chickens are usually reluctant to follow the shelter and therefore run the risk of being crushed, but after a few times they learn to walk along when the shelter is moved.

pastured/pasture raised \

Pasture confinement. \ Allowed access to fresh vegetative grazing much of the time. The meat of pastured broilers and the eggs of pastured layers contain less fat and more omega-3s and other nutrients than that of chickens not on pasture. Pasturing has the additional advantage of keeping chickens healthy by avoiding the buildup of pathogens that occurs around fixed housing; however, it requires enough good pasture (or unsprayed lawn) to move the shelter as often as necessary to prevent an accumulation of droppings for the health of the chickens and to avoid burning the grass with too much nitrogen-rich manure.
[Also called: day range]   See also: free range; range feeding

pasture management \ The

quality of a pasture determines how much of their nutritional needs chickens can obtain by foraging. Plant growth is generally poor during the heat of summer and the coldest months of winter, while growth is greatest during the temperate days of spring and fall. When pasture grows faster than the chickens can eat it, mowing may be necessary to keep plants in the vegetative, or growth, stage and thus easier to digest and more nutritious. Mowing also lets in sunlight to help minimize the buildup of infectious organisms.

Exactly how long they may be kept in one spot varies with the number, size, and activity level of the chickens and the seasonal growth of the pasture. Here are some guidelines:

- Move chickens in when the pasture is no more than 5 inches (12.5 cm) tall. If it gets taller, mow it.
- Move chickens out when the pasture has been grazed down to 1 inch (2.5 cm) or bare spots begin to appear.
- The longer chickens are kept in one place, the longer pasture plants take to recover after the shelter is moved.
- Avoid returning chickens to the same piece of ground twice within the same year.

Over the years, an accumulation of chicken droppings will cause pasture soil to increase in acidity. When soil pH drops below 5.5 as determined by a soil test, spread lime at the rate of 2 tons per acre (2 metric tons per 0.5 hectare). Then let the pasture rest to give plants time to rejuvenate and to break the cycle of parasitic worms and infectious diseases.

pathogen \ Any disease-producing

agent or organism, especially a microscopic organism.

pathogenic \ Capable of causing a

disease.

pathologist \ A veterinary profes-

sional who does a postmortem examination of internal damage caused by disease to determine the nature of the disease. Many diseases have such similar signs that the only way to get a positive diagnosis is by taking a few sick or recently dead birds to a pathologist for analysis. A pathologist offers a diagnosis but does not suggest how to treat any remaining sick chickens. For information on treatment, the path report must

P

pastured

be analyzed by a practicing veterinarian, a state Extension poultry veterinarian, or an Extension poultry specialist.

pathology lab \ A laboratory in which a pathologist works. Every state has one or more state-run path labs, sometimes called veterinary or animal disease diagnostic laboratories, and some state labs specialize in poultry pathology.

path report \ A formal written diagnosis prepared by a pathologist.

pearl eyed

pea comb

pea comb \ A low-growing comb consisting of three parallel ridges, which might be viewed as three small single combs joined together at the rear, with the center ridge being slightly higher than the outer ones and all three edged with tiny scallops. Pea comb is characteristic of these breeds: Ameraucana, Araucana, Aseel, Brahma, Buckeye, Cornish, Cubalaya, Shamo, Sumatra.
[Also called: triple comb]

pearl eyed \ Having creamy white or pale bluish-gray–colored eyes, characteristic of the Aseel, Cornish, Malay, and Shamo.
[Also called: daw eyed]

peck \ A sharp, quick stroke with the beak, usually with the intent of obtaining something to eat. Chickens also peck each other as a form of social control; for instance, a subordinate hen getting too close to a dominant hen at the feeder might get a harsh peck as a reminder to move away. Pecking may lead to picking, particularly among young growing chickens. *See also: pecking order; picking*

pecking order \ The social hierarchy that develops among a population of chickens and determines such things as which ones eat first or roost on the highest perch. Chicks start establishing their place in the pecking order at about six weeks of age. When both sexes are present, the peck order develops on three levels: among all the males, among all the females, and between the males and the females.

Cocks are usually at the top of the peck order, then hens, then cockerels, and finally pullets, although the cockerels will work their way up through the hens as they mature, and similarly, maturing pullets work their way up. A newly introduced chicken must also work its way up but won't necessarily start at the bottom.

Once the pecking order is established, a chicken of lower rank infringing on the space of one of higher rank will get a glare, and sometimes a harsh peck, from the higher-ranking bird. Fighting is thus kept to a minimum and mainly involves challenges to the top cock. The older he is, the more often he'll be challenged by younger upstarts.

Other results of the pecking order are that dominant cocks mate more often than lower-ranking cocks, but submissive hens mate more often than dominant hens because they are more easily intimidated and therefore crouch more readily. And in a flock of chickens with various comb styles, those with single combs are commonly higher in rank than those with other comb styles. **See box on page 200 for maintaining a stable pecking order.**

pediculosis \ Condition of being infested with lice.

PEARSON'S SQUARE

This is a simplified method for balancing rations by increasing or decreasing a specific nutrient, such as adjusting the protein in a ration by combining it with a supplement. To use Pearson's square, you need to know how much protein (to follow the example) you want to end up with and the protein content of both the existing ration and the supplement. To raise protein, choose a supplement that's higher in protein than the ration; to reduce protein, choose a supplement that's lower in protein than the ration.

Draw a square on a piece of paper. In the upper left corner, write the percentage of protein of the existing ration. In the lower left corner, write the percentage of protein contained in the supplement. At the center of the square, write the percentage of protein you want to end up with.

Working from the upper left toward the lower right (following the arrow in the illustration), subtract the smaller number from the larger number. Write the answer in the lower right corner. Moving from the lower left toward the upper right (again follow the arrow), subtract the smaller number from the larger number. Write the answer in the upper right corner. The number in the upper right corner tells you how many pounds of ration and the number in the lower right corner tells you how many pounds of supplement to mix together to achieve the desired amount of protein.

Use a spring scale for weighing feed, or weigh yourself on a bathroom scale holding an empty bucket, then add feed to the bucket until the total weight is increased by the amount needed.

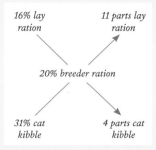

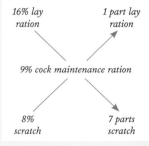

Here 16 percent layer ration is combined with 31 percent cat kibble to create a breeder-flock ration containing 20 percent protein. Since we want to increase the protein content, the number in the lower left corner is greater than the number in the upper left corner. Pearson's square suggests combining 11 pounds (5 kg) of layer ration with 4 pounds (1.8 kg) of kibble to get a 20 percent breeder ration.

Here 16 percent layer ration is combined with 8 percent scratch grain to create a 9 percent cock maintenance diet. Since we want to reduce the protein content, the number in the lower left corner must be less than the number in the upper left corner. Pearson's square suggests combining 1 pound (0.45 kg) of layer ration with 7 pounds (3.2 kg) of scratch to achieve a 9 percent cock maintenance ration.

pedigree \ A complete record of a chicken's ancestry. Chickens have no registry and therefore no registration papers. A breeder's pedigree (when available) is the only record of a chicken's ancestry. Most breeders who keep pedigree records do so to track various mating combinations so they can repeat the successful ones, and they guard their pedigrees as proprietary information.

Keeping track of each chick's ancestry requires banding every breeder with a unique identification code and marking every hatching egg with the identity of the mating that produced it, so each hatched chick may be traced back to a specific mating. When eggs from more than one mating are incubated together, the newly hatching chicks are kept separate by means of pedigree baskets.

pedigree basket \ A basket-type enclosure used in an incubator or hatcher to separate chicks resulting from different matings. Pedigree baskets may be homemade of hardware cloth or purchased as small plastic or wire baskets from an office-supply, school-supply, toy, drug-, or dollar-type store.

A reasonable size is whatever will hold the number of eggs you wish to put in it — typically one week's worth

MAINTAINING A STABLE PECKING ORDER

By governing a flock's social organization, a stable pecking order minimizes tension and stress among flock members. You can help your flock maintain a stable peck order in the following ways:

- Give your chickens plenty of room so the lowest-ranking birds can get away from those of higher rank.
- Design your facilities with enough variety so timid birds can find places to hide.
- Provide enough feeders and drinkers for the number of chickens you keep; otherwise, higher-ranking birds will chase away lower-ranking birds.
- If you have more than one cock, furnish one feeding station per cock and position feeders and waterers so no bird

has to travel more than 10 feet (3 m) to eat or drink.

- Avoid introducing new chickens into the flock, which causes a reshuffling of the peck order. Constantly disrupting the peck order by frequently introducing new chickens can lead to cannibalism.
- When introducing new birds, reduce bright lighting to make the unfamiliar chickens less conspicuous.
- If chickens constantly fight, look for management reasons such as poor nutrition, insufficient floor space, or inadequate ventilation.
- Never cull a chicken just because it is lowest in the pecking order — as long as you have at least two chickens, one will be lowest in rank.

of eggs from one hen — which may be determined by placing that number of eggs in a square on a flat surface and measuring how much space they take. To give the newly hatched chicks sufficient wiggle room, each basket should be big enough to hold two or three more eggs than you intend to put in it.

When using two or more identical baskets, keep track of which basket holds eggs/chicks from which mating by attaching a different colored or numbered leg band to each basket as a method of identification.

pedigree breeding \ Collecting
hatching eggs from specific matings, with emphasis on the characteristics of superior individuals, as compared to emphasizing a flock's overall production performance. Pedigree breeding is most often used for exhibition strains and to implement the recovery of endangered breeds. Pedigree breeding most commonly involves linebreeding, and its primary goal is progeny testing.

peduncle \ A small, stalklike
appendage consisting of skin-covered cartilage sprouting at the side of the head, below the ears, from which an ear tuft grows.

peep \ The sound a chick makes. A
chick peeps even before it hatches. After hatching, it makes a number of different peeping sounds by which you can tell whether or not it is content. The happy sounds tend to swing upward in pitch, while unhappy sounds descend in pitch. The most common peeps are as follows:

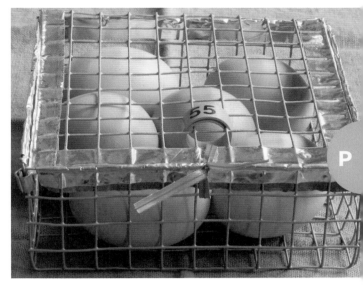

pedigree basket

- **PLEASURE PEEP** is a soft, irregular sound chicks use to maintain contact with each other and their mother
- **PLEASURE TRILL** is the soft, rapidly repeated sound chicks make when they've found food or are cozily nestled under the mama hen
- **DISTRESS PEEP** is a loud, sharp series of sounds made by chicks that are cold or hungry
- **PANIC PEEP** is a loud, penetrating peep of a chick that's scared or lost; it is similar to the distress peep, only louder and more insistent
- **FEAR TRILL** is the sharp, rapidly repeated sound made by a chick that sees something strange or potentially threatening
- **STARTLED PEEP** is the sharp, surprised cry of a chick that's been grabbed abruptly without warning

\ A newly hatched chick.

peepers \ Blinders. \ Newly hatched
chicks.

penciled

pellet \ A small, rod-shaped bit of compressed mash, designed so chickens can obtain a balanced diet by eating the entire pellet, rather than having the choice of picking out only what they like. The disadvantage to pellets is that they allow chickens to quickly satisfy their nutritional needs and then have little to do, and bored chickens tend to pick on one another.

pellet bedding \ Pine pellets sold as stall bedding. They are absorbent, odor free, less dusty than shavings, and longer lasting.

pelvic bones \ *See: pubic bones*

pen \ A small yard. \ A group of chickens housed together for breeding purposes, typically consisting of one cock and several hens. \ A cock (or cockerel) and four hens (or pullets) of the same breed and variety, entered into a show and judged together as a unit.

pen breeding \ Keeping a cock together with a small number of hens to collect their eggs for hatching. Pen breeding is similar to flock breeding except smaller numbers of hens are mated to only one cock at a time. Pen-bred birds may be pedigreed only on the cock's side unless each egg can be identified as coming from a specific hen. Positive identification of parentage is possible if each hen lays an egg of size, color, or shape that differs from the others. The only other way to be certain of each hen's identity is to use trap nests.

penciled, penciling \ A feather pattern consisting of narrow, continuous, sharply defined stripes of uniform width. In some varieties, such as penciled Hamburgs, penciling appears as crosswise stripes. In other varieties, such as silver penciled Wyandottes, the stripes follow the contour of the feather's edge.

Penedesenca \ A dual-purpose breed from the Penedès community in the Principality of Catalonia, Spain. The Penedesenca is closely related to the Empordanesa but is meatier. It has a carnation comb, has no bantam counterpart, and comes in solid black and a few color patterns, of which black is most common. The crele variety is autosexing — newly hatched cockerels are gray, pullets are brown. Like other Mediterranean breeds, Penedesencas are active and flighty. The

Penedesenca
hen

hens are good layers and, despite having white earlobes, produce eggs with dark chocolate–brown shells, sometimes speckled, and they seldom brood.

peppering \ An undesirable sprinkling of small dark dots over a light-colored feather, typically occurring as black spots on the tail feathers of buff varieties.

perch \ *See: roost*

performance bred \ A population of chickens, usually hybrid, developed for either a superior rate of laying or a rapid growth rate.
[Also called: production bred]

Persian Rumpless \ *See: Manx Rumpy*

persistency of lay \ The ability of a hen to lay steadily over a long period of time.

pet quality (PQ) \ A chicken that is not bred for high egg production or rapid growth as a broiler and does not conform closely enough to the standard description for its breed and variety to be of show quality. Pet-quality chickens are generally produced by chick mills looking to make a fast buck, but may also be lesser-quality offspring from a carefully bred flock.

Chickens kept as pets may be any breed with a calm disposition. Even a breed typically known to be flighty may become sociable if you start with chicks and spend a lot of time with them.

pH \ A numerical indication of the acidity or alkalinity of a substance. Pure water has a pH of 7; therefore, 7 is neutral, above 7 is alkaline, below 7 is acidic. The symbol pH stands for potential for hydrogen (H is the symbol for hydrogen), since the acidity or alkalinity of a substance is determined by how many hydrogen ions it forms in a specific volume of water. The importance of pH to a chicken keeper includes the following:

- Keeping chickens healthy
- Encouraging chickens to drink water
- Conditioning soil
- Maintaining shell quality in hot weather

See also: competitive exclusion; pasture management; probiotic; shell, thin; sour crop; vinegar

Penedesenca rooster

Phoenix
rooster

phosphorus \ A naturally occurring chemical element needed by hens to metabolize calcium. Without an adequate amount of phosphorus, calcium cannot be absorbed, and hens may experience calcium deficiency despite the availability of a calcium supplement. Sources of phosphorus include dicalcium phosphate, bone meal, and soft rock phosphate. The correct ratio of phosphorus to calcium is 1:2. When both supplements are offered separately and free choice, the hens will ingest the correct balance.

photostimulation \ *See: controlled lighting*

picking \ The harmful behavior of chickens that persistently peck at each other's flesh, feathers, or eggs.   *See also: cannibalism* \ To remove the feathers from a chicken killed for meat. Feathers may be removed by handpicking, wax picking, or machine picking. An alternative to picking feathers is to skin the chicken.
[*Also called: plucking*]

Phoenix \ A longtail breed originating in Japan and developed in Germany. It shares its history with the Yokohama and is similar in its small size and its conformation, differing primarily in having white earlobes and a single comb and coming mainly in color varieties common to the game breeds.

To distinguish it from the Yokohama, the Phoenix was given the name of a beautiful mythological bird and promoted as being a bird of fables and legends. The cock's tail can get pretty impressive, with saddle feathers up to a foot and a half long and sickles twice that length or more. To grow such long feathers, the Phoenix needs a high-protein diet.

The hens are poor layers of eggs with tinted shells, although their rate of lay may be improved with extra protein. They may brood and are protective mothers. The Phoenix comes in both large and bantam versions.

Phoenix hen

picking machine \ A mechanical device designed to remove feathers from chickens being butchered for meat. Picking machines come in two styles:

A TABLETOP PICKER has a rotating drum with rubber fingers against which one chicken is held at a time while the feathers are flailed off.

A TUB PICKER has rubber fingers lining a rotating drum into which one or more chickens are dropped; the feathers are flailed off as the drum spins.

picking wax \ Paraffin sold for the purpose of removing feathers from butchered poultry.   *See also: wax picking*

pickout \ A condition in which vent picking escalates to the point of causing a gaping wound through which the picked chicken's internal organs are pulled out.   *See also: blowout; cannibalism; vent picking*

pied \ Description of plumage having feathers of more than one color or more than one shade of a single color.
[Also called: particolored]

pin bones \ *See: pubic bones*

pinfeathers \ The pinlike tips of newly emerging feathers appearing when a chick first feathers out and again whenever it undergoes a molt. Pinfeathers are of significance primarily because they can invite cannibalistic picking and because they make meat birds more difficult to pluck.

Pinfeathers contain a supply of blood to nourish the growing feather (hence their other name, blood feathers) and therefore invite cannibalistic picking, especially around the tail and along the back, at a time when molting chickens crave additional protein. Once the feathers are fully formed, the blood supply is cut off, reducing the desire of chickens to pick one another's feathers.

Pinfeathers on meat birds won't come out cleanly no matter how carefully the chickens are plucked. Either skin the chickens instead of plucking them, or hold off butchering for a few more days while the feathers grow past the pinfeather stage. The few remaining pinfeathers may be removed with a pinning knife.
[Also called: blood feathers]

pinion \ The last segment at the end of a chicken's wing. \ To surgically remove the last segment of one wing for the purpose of permanently restricting flight.

pinion feathers \ *See: primary feathers/primaries*

pinning knife \ A pinfeather scraper with a rounded 3- to 4-inch (7.5 to 10 cm) blade, used to clean the skin of chickens being butchered for meat. After the main feathers have been removed, any remaining pinfeathers may be squeezed out by holding the bird under a running faucet and scraping the skin with the pinning knife. Stubborn pins may be pulled out between the knife blade and your thumb. A suitable substitute for a pinning knife is a paring knife with a dull blade.

pip

white offers camouflage protection for foragers and setters. For chicken keepers plumage color is important as an indication of breed purity. Slight variations among family lines sometimes help identify the strain.

plumage color

laced

splashed

cuckoo

pip \ The hole a newly formed chick makes in its shell when it is ready to hatch. \ The act of making the hole.

Pit Game \ Early name for Old English Game.

pitting \ Cockfighting.

plucking \ *See: picking*

plumage \ All the feathers covering a chicken, from the Latin word *pluma*, meaning feather.

plumage color \ Chicken breeds and varieties come in hundreds of colors and color patterns ranging from a rainbow of single colors such as red, white, or blue to patterns such as speckled, barred, or laced. Most strains of commercially produced chickens are white, whether bred for the production of meat or eggs. The greatest variety in color occurs among exhibition breeds and varieties.

The importance of plumage color to chickens is that any color other than

Plymouth Rock \ A large, meaty dual-purpose breed developed in Massachusetts and named after the boulder on which Pilgrims supposedly first landed in Plymouth, Massachusetts. The Plymouth Rock is a hardy, docile breed with a single comb. It may be large or bantam in size and comes in a few color varieties. Barred is the original variety and remains the most popular for backyard flocks, while white Rocks are used in producing industrial hybrid broilers. The hens are excellent layers of brown-shell eggs and are also excellent broodies.

pneumatic bone \ One of several hollow air-filled bones that aid a chicken's respiration and also lighten its weight to help it fly.

pneumonia \ Any disease of the lungs.

point of lay (POL) \ The age at which a pullet is due to start laying.

points \ The tips of the serrations at the top of a single comb. The front and back points tend to be smaller than those in the center, and the total number of points varies with breed. \ Weighted values assigned to, or deducted from, specific traits in the evaluation of a chicken competing in an exhibition.   *See: general scale of points*
See page 66 for illustration.

poison \ Chickens are susceptible to being poisoned by pesticides, herbicides, rodenticides, fungicide-treated seed (intended for planting), wood preservatives, rock salt, antifreeze, and the excessive use of disinfectants. Hens can be poisoned by cedar shavings or mothballs inappropriately placed in nests to repel

Plymouth Rock hen, *barred plumage*

Plymouth Rock rooster, *Columbian plumage*

lice and mites. Pastured poultry can be poisoned by toxic plants.

Baby chicks are more susceptible than mature chickens to toxins in their environment. They can die from carbon monoxide poisoning while being transported in the poorly ventilated trunk of a car. They can become sick from the overuse of phenol disinfectants, especially in an inadequately ventilated brooder. They can obtain a deadly dose of a coccidiostat (nicarbazin, monensin, sulfaquinoxaline) added to water in warm weather, when they drink more than usual. Despite all these possibilities, poisoning of chicks or grown chickens is fairly uncommon.　*See also: toxic plants*

Polish \ An old breed developed in the Netherlands as a layer and now

Polish hen preening

considered primarily ornamental. Why they are called Polish is a matter of much conjecture. Theories include the following:

- The region of Eastern Europe where the breed originated was then part of Poland
- The topknot resembles feathered caps once worn by Polish soldiers
- The name is a corruption of one of several villages with names starting with P, through which chickens were transported on their way to Holland
- The name derives from the Middle Dutch word *pol*, meaning head or top

Polish chickens are distinguished by their opulent crest, or topknot, which hangs over their eyes and interferes with their ability to see. They have a V-comb or no comb at all, may be large or bantam, come in several color varieties, and may be found with or without a beard and muffs. In the Netherlands, the bearded and nonbearded varieties are considered to be two different breeds. The hens are good layers of white-shell eggs and seldom brood.

poop \ *See: droppings*

pope's nose \ *See: uropygium*

popeye \ The emaciation of chicks as a result of disease, nutritional deficiency, or starve-out. It is called popeye because the chicks' eyes appear disproportionately large in relation to their wasted bodies.

pop hole \ A chicken-size door cut into the wall of the shelter. A good-size single-chicken pop hole is 10 inches wide by 13 inches high (25 by 33 cm); for really large chickens 12 by 14 (30 by 35 cm) would be better. When an automatic door closer is used, the pop hole must be sized to fit. If an automatic closer is not used, the pop-hole cutout may be used as a door to latch the pop hole shut at night to keep out predators. The pop-hole door may be either hinged at the bottom and opened to create a ramp or hinged at the top or side and fastened against the wall during the day.

Chickens love to sit on the pop-hole ledge on their way in or out. When one chicken is there, no others can get in or out. A range or pasture shelter therefore needs a wide pop hole so foraging chickens can dash inside in the event of an aerial attack.
[Also called: poultry portal]

portable shelter \ A chicken coop that is periodically moved to fresh ground. A portable shelter either has a floor, in which case it must be used in combination with a movable fence to keep the chickens close by, or it has no floor and must be moved often. To aid in moving, portable shelters come in four basic styles:

- With handles, to be carried by two people
- On skids, to be moved by hand if light enough; otherwise by ATV, truck, tractor, or draft animal
- On wheels without an axle, to be moved by hand; the wheels may be incorporated into the shelter design or on a separate dolly
- On wheels with an axle to be moved by ATV, truck, tractor, or draft animal

See also: pasture confinement

post \ To conduct a postmortem examination.

postmortem \ The examination of a dead chicken to determine the cause of death, or of a recently killed sick chicken to determine what was making it sick, in an attempt to prevent the spread of disease to others in the flock. From the Latin *post*, meaning after, and *mortem*, meaning death.

poultry \ Chickens and other domesticated barnyard birds raised for eggs, meat, or fun. From Old French *poulet*, meaning young fowl.
See also: fowl

pop hole

poultry judge \ A person licensed by the American Bantam Association, American Poultry Association, or both as being qualified to render an opinion on the quality of chickens entered into a competition insofar as each conforms to the judge's interpretation of the standard description for the breed and variety.

poultry netting/wire \ *See: hexagonal netting/wire*

poultry portal \ *See: pop hole*

precocial \ Capable of independent activity, including self-feeding, almost from the moment of birth. From the Latin word *praecox*, meaning mature before its time. The chief characteristic of precocial chicks is their spryness soon after hatching, allowing them to be separated easily from the mother hen. A resulting characteristic is that they communicate with the mother hen before they hatch: chicks in the shell peep, and the broody hen clucks. Chicks thus learn to recognize their mother's voice while still in the shell and enter the world with the ability to quickly find the hen should they begin to stray — a trait essential to their survival.

predator \ One animal that hunts another for food or fun. Chickens fall prey to a large number of flying and four-legged predators, each of which leaves its own calling card that helps in identifying the culprit.

predator control \ Predation may be controlled in one of two ways — eliminate the predator's point of entry or eliminate the predator. The latter usually involves either trapping or shooting. For a species that's protected by law, the only option is to eliminate point of entry.

The easiest way to eliminate point of entry is to confine chickens indoors, if not all the time, at least at night. An automatic door is handy for closing the pop hole at night and opening it in the morning. A bright security light discourages some predators and helps you see when you do have a nighttime problem.

Covering window and vent openings with ½-inch (1.5 cm) hardware cloth keeps out the smaller predators and prevents larger ones from ripping through screens or chicken wire. Diggers may be discouraged by a deep concrete foundation around the shelter and an apron around the fenced yard. A sturdy close-mesh fence keeps out many predators, and scare wires help keep out the rest. To deter flying predators, cover a small run with wire mesh and crisscross a larger yard with wires strung 7 feet (2 m) off the ground. From these wires hang fluttering CDs, DVDs, or strips of old sheet.

Move pastured poultry often. Mow down tall grass, weeds, and brush. Keep shelters away from trees and fence posts. As additional protection install a guardian animal, such as a dog or burro. To deter night-time predators, install a few solar-powered night predator lights that blink regularly between dusk and dawn — the small red flash appears to be the eye of another animal on the prowl, making night-time predators feel threatened because they fear being watched.

preen gland \ *See: uropygial gland*

P

PREDATOR IDENTIFICATION

BIRDS MAULED

Clues	Likely time	Predator
Multiple birds mauled	Any time	DOG
Pullet wounded around vent area	Day	CHICKENS (PICKOUT)
Hen with slice wounds along sides of the back	Day	ROOSTERS (TREADING)
Growing birds with missing toes or wounds near top of the tail	Day	CHICKENS (CANNIBALISM)
Growing birds with bites on breast or thigh	Night	OPOSSUM
Growing birds with bites around hock	Night	RAT

SEVERAL BIRDS KILLED

Clues	Likely Time	Predator
Birds mauled, but not eaten; fence or building torn into; feet pulled through cage bottom and bitten off	Any time	DOG
Bodies neatly piled, killed by small bites on neck and body	Night	MINK
Birds killed by small bites on neck and body, bruises on head and under wings, bodies neatly piled, faint skunk odor	Night	WEASEL
Rear end bitten, intestines pulled out; bodies stashed and eaten later	Night	FISHER, MARTEN
Bloodied bodies surrounded by scattered feathers	Night	FERRET, FISHER, MARTEN, MINK
Chicks dead; may be faint lingering odor	Night	SKUNK
Back of heads and necks eaten	Night	MINK, WEASEL
Heads and crops eaten	Every 5–7 nights	RACCOON

chart continues on following page

PREDATOR IDENTIFICATION

ONE OR TWO BIRDS KILLED

Clues	Likely time	Predator
Entire chicken eaten on site	Dusk or dawn	HAWK
Abdomen eaten; entire bird eaten on site	Night	OPOSSUM
Deep marks on head and neck, or head and neck eaten; may be feathers around fence post	Night	OWL
Entire chicken eaten or missing; may be scattered feathers	Early morning	COYOTE
One bird gone; may be scattered feathers	Dusk or dawn	FOX
Meatier parts eaten, other parts scattered, skin and feathers remain	Night	DOMESTIC CAT (RARE)
Chicks pulled into fence, wings and feet not eaten	Nightly	DOMESTIC CAT
Chicks killed, abdomen eaten (but not muscles and skin); may be lingering smell	Night	SKUNK
Head bitten off; claw marks on neck, back, and sides; body partially covered with litter	Night	BOBCAT (RARE)
Bruises and bites on legs; partially eaten chick or young chicken with head down tunnel	Night	RAT
Backs bitten, heads missing, necks and breasts torn, breasts and entrails eaten; bird pulled into fence and partially eaten; body found away from shelter; may be scattered feathers	Every 5–7 nights	RACCOON

ONE BIRD MISSING

Clues	Likely Time	Predator
Ranged bird missing, feathers scattered or no clues	Dusk or dawn	FOX
A few scattered feathers or no clues	Dusk or dawn	HAWK
Fence or building torn into, feathers scattered	Any time	DOG

ONE BIRD MISSING, *continued*

Clues	Likely Time	Predator
Ranged bird missing, feathers scattered or no clues	Dusk or dawn	**COUGAR, CATAMOUNT, MOUNTAIN LION, PUMA, PANTHER** (RARE)
A few scattered feathers or no clues	Night	**OWL**
Small bird missing, lingering musky odor	Night	**MINK**
Small bird missing, wings and feathers remain	Night	**DOMESTIC CAT**
Ranged bird missing, no clues	Night	**BOBCAT** (RARE)

SEVERAL BIRDS MISSING

Clues	Likely Time	Predator
No clues	Any time	**HUMAN**
Ranged birds missing, feathers scattered or no clues	Dusk or dawn	**FOX**
Ranged birds missing, no clues	Early morning	**COYOTE**
Ranged birds missing, no clues	Day	**HAWK**
Chicks missing, no clues	Day	**SNAKE**
Small birds missing, bits of coarse fur at shelter openings	Night	**RACCOON**
Chicks or young birds missing, no clues	Night	**RAT, DOMESTIC CAT**

EGGS MISSING FROM NEST

No clues	Day	**SNAKE**
Empty shells in and around nests	Any time	**DOG**
Empty shells in nest or near housing	Day	**JAY, CROW**
No clues	Night	**RAT**
No clues or empty shells in and around nests; may be faint lingering odor	Night	**SKUNK**
Empty shells in and around nests	Night	**RACCOON, MINK**
Empty shells in and around nests	Nightly	**OPOSSUM**

preening \ Part of a chicken's typical grooming behavior, engaged in following a dust bath and at other times. Preening involves straightening the feathers with the beak and rubbing the head against the preen gland, then against the feathers to oil them with preen oil. Preening cleans the feathers of dead skin and parasites, distributes oil through the feathers, and maintains proper feather structure.

preen oil \ Oil secreted by the uropygial gland and spread on feathers during grooming. Preen oil conditions the feathers and keeps them from becoming brittle.

premium \ A prize given at a poultry show. The value of the prizes often determines the amount of competition expected in a given class. Competition is generally stiffest in a class involving a sanctioned meet, at which an organization furnishes awards for its members.

premium list \ A document listing the classes and varieties that will be accepted for exhibition at a particular show. The premium list should indicate which breed standard (American Poultry Association or American Bantam Association) will be used, as the two don't always agree. Other information found in a premium list includes:

- The show's time, date, and location
- The deadline for entering
- Entry fees involved
- Health requirements
- The days and times by which birds must be cooped in and cooped out

- Acceptable methods for identifying each bird
- Premiums being offered

prepotent \ The ability of a chicken to pass on hereditary attributes to the majority of its offspring. Prepotency results from homozygosity.

primary coverts \ The contour feathers covering the bases of the wing's primary feathers.

primary feathers/primaries \ The 10 largest feathers of the wing. These long, stiff feathers, which grow from the pinion, run from the tip of the wing to the short axial feather that separates the primaries from the secondary feathers. When the wing is folded at rest, the primaries are completely covered by the secondaries.
[Also called: flight feathers; flights; pinion feathers]　See also: *flight feathers/flights*
See page 115 for illustration.

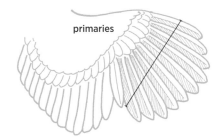

primaries

probiotic \ A live microbial supplement that stimulates the growth of beneficial microorganisms in the intestines. The small intestine of a healthy chicken (and also a healthy human) is populated with a number of beneficial bacteria and yeasts, called intestinal microflora or gut flora, that aid digestion and also produce antibacterial compounds and enzymes that stimulate the immune system. If for any reason these beneficial microbes get out of balance, disease-causing microbes take over and cause an enteric disease. Adding apple cider vinegar to the drinking water helps encourage microflora to flourish.

A chick acquires some microflora through the egg and gains more from the environment, particularly from properly composting litter. Microflora are naturally present in certain foods, including grains, meats, and fermented milk such as yogurt and kefir. Chickens that eat a varied diet or are free to peck out some of their sustenance from the environment typically do not need a probiotic.

Antibiotics and other antimicrobials kill both disease-causing microbes and beneficial microflora. Any chicken that has been treated with an antibiotic or subjected to extreme stress may benefit from a probiotic.

Chicks raised in a brooder acquire beneficial gut flora more slowly than chicks raised under a hen. To enhance their immunity, a probiotic is either dissolved in water or sprinkled on feed to give chicks an early dose of the same gut flora that will eventually colonize their intestines. A handy substitute for this supplement is a little live-culture yogurt, although too much yogurt will cause diarrhea. *See also: vinegar*

processor \ A person or firm that kills, cleans, and packages chickens grown for meat.

producer \ A person or firm that raises meat birds or laying hens.

production black \ *See: California Gray*

production bred \ A population of chickens, usually hybrid, developed for either a superior rate of laying or a rapid growth rate.
[Also called: performance bred]

production red \ A hybrid cross between Rhode Island Red and New Hampshire, bred as a superior layer of brown-shell eggs.

progeny testing \ Measuring the value of a breeder by the quality of its offspring. Progeny testing is a long-term process involving keeping track of all birds from each mating, not just the occasional superior individual. Certain matings may produce outstanding traits in one area, while others produce outstanding traits in another area. Producing superior birds combining both traits sometimes requires setting up a separate breeding line for each.

A particular individual or mating that produces superior offspring should be used as often as possible, while matings that produce unhealthy or otherwise substandard chicks should be avoided — unless you are engaged in a recovery program that requires preserving the greatest possible degree of genetic diversity, in which case retain as future breeders the best offspring resulting from each mating.

prolapse \ A natural process by which eggs are laid. The oviduct's shell gland, or uterus, pushes a completed egg through the cloaca by holding the egg tightly and prolapsing, or turning itself inside out, to deposit the egg outside the vent, then withdraws back inside the hen. If an egg is too big, or a pullet is immature when laying begins, the uterus may not retract back inside. Uterine tissue that protrudes outside the vent and remains prolapsed is a serious condition called blowout.
[Also called: eversion]

protein \ A nutrient furnished by any feedstuff that is high in amino acids. Chickens need a lot of protein, especially chicks that are growing, hens that are laying, roosters that are expected to fertilize eggs, or any chicken that's going

through a molt. Furnishing quality protein is the most expensive part of feeding chickens.

Examples of feedstuffs containing protein include eggs, meat, milk, nuts, seed germs, and soybeans and other legumes. Chickens of different ages and levels of production need different levels of dietary protein. Broilers require the greatest amount (20 to 24 percent), while mature cocks outside breeding season require the least (9 percent). Whenever the protein in a flock's diet is adjusted by more than a percentage point or two, the change must be made gradually to avoid intestinal upset and diarrhea.

Chickens eat to meet their energy needs and require more energy (carbohydrates) to stay warm in cold weather. Since energy is less expensive than protein, you can save money by reducing protein (increasing carbohydrates) during winter. On the other hand, a chicken's need for dietary protein increases during the annual molt, since feathers are 85 percent protein. A little supplemental animal protein will help them through the molt, as well as improve the plumage of show birds. Compared to the protein in grains, animal protein is rich in the amino acids a chicken needs during the molt.

protozoan (plural: protozoa)

\ A single-cell microscopic animal that may be either parasitic or beneficial. Some protozoa make up part of the gut flora. Others cause serious diseases, including coccidiosis.

prolapse ⟶

PROTEIN REQUIREMENTS

Type	Age	Ration	% Protein
BROILERS	0–3 weeks	Broiler starter	20–24
	3 weeks–butcher	Broiler finisher	16–20
LAYERS	0–6 weeks	Pullet starter	18–20
	6–14 weeks	Pullet grower	16–18
	14–20 weeks	Pullet developer	14–16
	20+ weeks*	Layer	16–18
COCKS	Maintenance	Layer + scratch	9
BREEDERS	20+ weeks*	Breeder	18–20

*Layer or breeder ration should not be fed to pullets until they start laying at 18 to 20 weeks for Leghorn-type hens, 22 to 24 weeks for other breeds.

From: *Storey's Guide to Raising Chickens*, 3rd edition

SOURCES OF ANIMAL PROTEIN

Animal protein can come from any of the following:

- High-quality cat food (not dog food, most of which derives protein from grains)
- Raw meat from a reliable source (not chicken; feeding an animal the meat of its own species can perpetuate diseases)
- Fish (but don't feed fish to a chicken you plan to eat anytime soon, or the meat may taste fishy)
- Molting food, sold by pet stores for caged songbirds (it's expensive but lets you circumvent issues with potentially toxic pet foods and bacteria-laden meats)
- Scrambled or hard-boiled eggs, mashed
- Sprouted grains and seeds, particularly alfalfa and sesame seeds (sprouting improves protein quantity and quality)
- Mealworms
- Earthworms

proventriculus \ A chicken's true stomach, where enzymes break down partially digested feed passed down from the crop.

pubic bones \ A pair of sharp, slender, pointed bones positioned between the keel and the vent. The flexibility of these bones offers a reliable way to evaluate a hen's egg production.
[Also called: pelvic bones; pin bones]
See also: layer

pullet \ A female chicken that is less than one year old. From the French word *poulet*, meaning little hen.

pullet breeder \ A cock or hen selected to produce outstanding standard-bred pullets for exhibition. For some varieties, achieving exhibition quality requires breeding different strains to obtain females and males that conform to the standard for their variety. The partridge Wyandotte is one such variety. To obtain exhibition-quality pullets with fine penciling requires using pullet-breeding males that display the requisite penciling, but instead of having the solid black breast of an exhibition-quality cock, pullet breeders have patches of red on their breasts.   *See also: cock/cockerel breeder*

pulley bone \ *See: wishbone*

Punnett Square \ A diagram, named after the British geneticist Reginald Punnett, used to show the probable genetic makeup and appearance of the offspring of a hybrid cross when the parents' genes are known. As a simple example, slow feathering is caused by a dominant allele (K), while fast feathering is caused by a corresponding recessive allele (k). A slow-feathering hen therefore has the dominant form of the gene on her Z sex chromosome ($Z^K W$), and a fast-feathering cock has the recessive gene on both sex chromosomes ($Z^k Z^k$). A Punnett square shows that if you mate a slow-feathering hen with a fast-feathering cock, all the resulting cockerels will inherit one dominant and one recessive gene ($Z^k Z^K$) and therefore will be slow feathering; all the pullets will inherit the recessive gene ($Z^k W$) and therefore will be fast feathering.   *See also: feather sexing*

Punnett Square

SLOW-FEATHERING HEN	FAST-FEATHERING COCK		OFFSPRING
	Z^k	Z^k	
Z^K	$Z^k Z^K$	$Z^k Z^K$	Slow-feathering cockerels
W	$Z^k W$	$Z^k W$	Fast-feathering pullets

purebred \ The offspring of a hen and a rooster of the same breed and variety. Whether or not these chickens may be called purebred is a matter of contention. Some people argue they cannot be called purebred because chickens have no registry and therefore no papers as proof of lineage. *See also: straightbred*

pygostyle \ *See: tailbone*

Pyncheon bantam \ An old and rare breed of true bantam of unknown origin, perhaps coming from Belgium, where the same color pattern was developed in other breeds. The name Pyncheon likely derives from the family name of the originator or a past breeder. This chicken has a single comb backed by a tassel, or small tuft of feathers pointing rearward. The original color pattern is mille fleur, although related color varieties have been developed. These bantams are docile and friendly and are capable fliers. The hens are fair layers of eggs with tinted shells and make good broodies, although fertility tends to be low.

P

Pyncheon bantam hen

Q R

quality \ The degree of excellence of show stock, which may be described as show quality (SQ) or not show quality (NSQ), of which the latter may be further described as being breeder quality (BQ), or pet quality (PQ).

quartet \ One cock mated with three hens of the same breed and variety.
[Also called: breeding quartet]

quill \ The stiff, hollow, transparent base of a feather shaft where the feather attaches to the chicken's body. \ A quill feather.

quill feather \ Any large, stiff feather of the wing or main tail.
See page 114 for illustration.

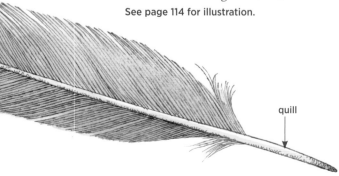

quill

radio frequency identification (RFID) \ A band or other object that uses radio waves to identify and track a chicken (or other animal or human) by means of a scanner or reader. Radio frequency identification devices are incorporated into both leg and wing bands but are rarely used by backyard breeders because readers are expensive and not all readers work with all brands of band.

rales \ Any abnormal sound coming from a chicken's lungs when the bird breathes in and out. Rales indicate the chicken is suffering from a respiratory disease.

range \ An open, fenced area on which chickens forage.

range feeding \ Similar to pasture confinement, except the chickens are allowed to come and go freely from their shelter to graze grass or pasture. The extra activity creates darker, firmer, more flavorful meat but also causes birds to eat more total ration because they take longer to reach target weight. Not everyone wants to raise broilers an extra few weeks on pasture, and not everyone appreciates the full flavor and firm texture of naturally grown chicken. As a result, backyard pasturing is less often used for growing meat birds than for keeping laying hens.

The time-honored method of range-feeding chickens, widely practiced in the days before confinement became conventional, is traditionally called free ranging. But the term "free range" has been corrupted by the USDA to mean "the poultry has been allowed access to the outside" (but does not necessarily require the birds to actually go outside), so keepers now use other descriptive phrases, such as range feeding.
[Also called: day range] *See also: free range; pastured/pasture raised*

range shelter \ A portable shelter for pasturing poultry. \ A rudimentary structure, typically consisting of short posts supporting a low flat or sloped roof, under which foraging chickens can find protection from sun, wind, and rain. In warm weather, a range shelter also should include a source of drinking water. Where chickens forage in an open area that extends 100 feet (30 m) or more from their housing, an arrangement of range shelters placed every 60 feet (18 m) or so encourages them to forage farther from their housing than they might otherwise.

rare breed \ A breed that is not commonly used in modern agriculture but once may have been. Various organizations work to conserve these breeds

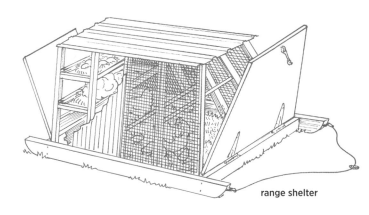

range shelter

to preserve their unique traits and safeguard genetic diversity. Each group defines the term "rare breed" in a slightly different way and lists breeds that best fit the definition. These organizations include the American Livestock Breeds Conservancy, Rare Breeds Canada, and the Society for the Preservation of Poultry Antiquities. *See also: endangered breeds*

Rare Breeds Canada (RBC) \ A Canadian organization working to conserve, monitor, and promote heritage and rare Canadian breeds.

rate of lay \ See box on page 222.

ration \ The main constituent in a chicken's daily diet. A commercial ration is designed to be used as a complete ration, although a lot of backyard keepers supplement commercial rations with table scraps, sprouts, garden gleanings, and similar foodstuffs to give their chickens a fresh and varied diet and/or minimize the expense of buying a purchased ration. *See also: commercial ration*

RATE OF LAY

Rate of lay is the number of eggs laid by a hen within a given time period, usually a year. A commercial-strain Leghorn averages between 250 and 280 white-shell eggs during the first year. A commercial hybrid brown-egg layer lays somewhat fewer eggs; red sex-links average 250 to 260 eggs per year, while black sex-links average about 240 eggs per year, but their eggs are larger than those of the red sex-link. Noncommercial breeds and strains lay somewhat fewer eggs, and some exhibition strains barely lay at all.

A healthy hen should lay for a good 10 to 12 years. Occasionally, you'll hear of a hen laying to the ripe old age of 20, by which time she'd be doing well to pump out an egg a week. Most layers don't live more than a few years, for reasons largely to do with either predators or owners who become dissatisfied with a low rate of lay. Aside from a gradual reduction due to age, a decrease in rate of lay is often the first general sign of disease.

In addition to a hen's breed and age, rate of lay is affected by external factors including temperature and light. Hens lay best when the temperature is between 45 and 80°F (7 and 27°C). When temperatures get much colder or much warmer, production slows. Most hens stop laying in winter, not because the weather turns cold, but because daylight hours are shorter in winter than in summer. When the number of daylight hours falls below 14, without controlled lighting hens may stop laying until spring.

An indication of a hen's worth as a breeder is the average rate of lay of all the hens in her family line. If the family, in general, consists of good layers, the hen is likely to pass the ability to her offspring, even if she herself is not a particularly outstanding layer. Conversely, an outstanding layer is not likely to pass her ability to her offspring if her rate of production is not typical of her family's average. ***See also: laying breed***

Redcap hen

ration balancing \ Adjusting the nutrients in a ration according to a flock's needs. The main cause of poorly balanced rations is feeding too much scratch grain or other carbohydrate-laden treats. *See also: feed (and entries following it for specific feeder ration recommendations)*

rats \ *See: rodents*

rattle \ Any abnormal sound coming from a chicken's throat when the bird breathes in and out. Rattling may indicate the chicken is suffering from a respiratory disease, or it may be a sign of gapeworm.

recessive trait \ A trait that
appears only when the gene governing that trait has two identical alleles for the trait. Most lethal genes are recessive. Examples of recessive traits include four toes, silkiness, single comb, stubs, vulture hocks, wry tail, and yellow skin.

red bird \ *See: red skin*

Redcap \ An old dual-purpose breed
originating in England and named after the massive red rose comb that covers the cock's head. The comb and color pattern do not fully develop until these chickens are two to three years old. Redcaps come in a single color pattern, red trimmed with black, and may be large or bantam. They tend to be active and flighty, are capable fliers, and are self-sufficient foragers. The hens are good, persistent layers of eggs with white shells and seldom brood.

red mite \ An external parasite
that feeds on chickens, and sometimes humans, but needs an avian host to reproduce. Red mites are active in warm climates during summer and are especially attracted to broody hens. They invade chickens primarily at night, when they appear as tiny red or black specks on a bird's body. After feeding they crawl along roosts to find a place to hide during the day. Adult mites can live for many months in unoccupied nests or housing. Control red mites by thoroughly cleaning the coop and dusting every crack and crevice with an insecticide approved for poultry.
[Also called: chicken mite]

Red Rock \ Trade name for black
sex-link hybrid chickens. Sometimes incorrectly designated as Rock Red.

red sex-links

red sex-link \ The
offspring of a red cock — a New Hampshire or Rhode Island Red — mated with a silver-based hen (Delaware, white Leghorn, white Rock, Rhode Island White, or silver laced Wyandotte). The chicks may be sexed easily at hatch because the males are white (although some may have a few black feathers), while the females are either buff or red. Trade names for red sex-link hybrids include Cinnamon Queen, Golden Comet, Gold Star, and Red Star.

cockerels (male)   pullets (female)

Redcap rooster

red skin \ A plucked chicken with skin that is bright red instead of the normal color. A red skin occurs when a chicken is scalded without first being properly bled.
[Also called: red bird]

Red Star \ Trade name for red sex-link hybrid chickens.

red worm \ See: gapeworm

replacement flock \ A group of pullets brought in to replace a group of older hens. Pullets generally reach peak production at 30 to 34 weeks of age. From then on laying declines approximately ½ percent per week until the hens molt or are replaced, typically by the time they reach 72 weeks of age. Keeping older layers has several disadvantages:

- As hens get older and their production declines, the shells of their eggs get rougher and weaker, the whites get thinner, and the yolk membranes become so weak they break when eggs are opened into a pan.
- As laying declines, the cost of feeding older hens becomes greater than the value of their eggs.
- If you're selling eggs, production by older hens may fall below the numbers you need to satisfy your customers.
- One-year-old layers may be sold to someone with less demanding needs, but by the end of their second year of laying, hens lose nearly all their value as layers.

- The older a chicken gets, the more likely it is to experience health complications.

The best time to replace layers depends, in part, on your chosen breed. Commercially developed hybrids still produce fairly well during their second year, while straightbred hens may peter out more quickly. Still, breeds that are better known for eggs than for meat may produce at a low but steady rate for years. Dual-purpose hens, on the other hand, tend to run to fat as they age, and fat hens do not lay well.

reportable disease \ A disease so serious it must, by law, be reported to a state or federal veterinarian. Diseases that are reportable in all states include avian influenza H5N1 and exotic Newcastle disease. States with a strong economic interest in the poultry industry may list additional reportable diseases. Since these lists change periodically, the best way to obtain a current list is to contact your county Extension agent or state poultry specialist.
[Also called: notifiable disease]

respiration rate \ The number of cycles per minute by which air is moved into and out of the respiratory system. A hen at rest breathes 31 to 37 times per minute, and a cock 18 to 21 times per minute. By comparison the normal rate of an adult human at rest ranges between 12 and 20.

respiratory disease \ Any of a number of diseases that invade a chicken's breathing apparatus. Chickens are particularly susceptible to respiratory illness, as their environment typically is high in humidity, dust, and ammonia fumes.

Respiratory diseases are characterized by labored breathing, coughing, sneezing, sniffling, gasping, and runny eyes and nose.

restricted egg \ A check, dirty, leaker, or otherwise inedible egg.

restricted feeding \ Feeding chickens a little at a time so they clean it all up fairly rapidly; the opposite of free-choice feeding. Restricted feeding is done for the following reasons:

- To reduce the protein intake of growing pullets to keep them from maturing too rapidly, because early laying results in fewer eggs of smaller size
- To reduce the feed intake of older, lightweight hens and breeders in the dual-purpose or meat categories to keep them from getting fat and lazy
- To encourage pastured hens to spend more time foraging
- To reduce the growth rate of Cornish-cross broilers so they won't grow so fast their little legs can't carry them
- To habituate exhibition birds to humans so they'll look forward to being visited

Restricted feeding requires having enough feeder space for all birds to eat at the same time. A typical regime involves offering as much feed as the chickens will clean up within 15 minutes, twice a day. For broilers, feed is typically removed overnight and offered free choice during the day. When broilers are pastured, this regime not only helps minimize lameness issues but ensures the birds will eagerly graze when moved each morning to new ground before being given a prepared ration.

A restricted-feeding program can cause chickens lowest in the peck order to not get enough to eat. And because chickens spend less time eating, if they are confined with little else to do, they tend to get bored and nervous and may pick on one another.

Rhode Island Red (RIR) \ A dual-purpose breed developed in the state of Rhode Island and subsequently designated the state's official bird. It comes in two comb varieties, single and rose, and one color pattern — dark red with a black tail — and may be large or bantam. The hens are among the best layers for a heavy breed, producing large eggs with brown shells. Among old strains, hens will go broody and make good mothers; among strains developed for increased egg production, hens seldom brood.

R

Rhode Island Red hen

Rhode Island White \ A dual-purpose breed developed in Rhode Island, unrelated to the Rhode Island Red but similar in conformation. Red sex-link hybrid layers may be produced by mating Rhode Island Red cocks to Rhode Island White hens. The Rhode Island White has a rose comb, although occasionally throws chicks with single combs; comes in a single solid color, white; and may be large or bantam in size. The hens are good layers of brown-shell eggs and seldom brood.

riboflavin \ Vitamin B_2, one of the vitamins often deficient in a poultry ration that has gone stale. The result of deficiency in a breeder flock is embryo deaths in early or mid-incubation of their eggs, depending on the degree of deficiency. Chicks that do hatch from deficient eggs or that are fed a deficient ration develop curled toes and may grow slowly. The damage is permanent when the deficiency is not immediately corrected. Riboflavin comes from leafy greens, milk products, liver, and yeast.

rigor mortis \ Stiffness following death and the cause of toughness in a plucked chicken that has not been sufficiently aged before being cooked. *See also: aging (of meat)*

roach back \ A deformity in which the back is humped. Roach back is not the same as the convex back characteristic of a Malay or the heavy hip muscles that give the Cornish cock's back a roundish look.

roaster \ A cockerel or pullet, usually weighing 4 to 6 pounds (1.8 to 2.7 kg), suitable for cooking whole in the oven.

Rock Cornish \ The term used by the United States Department of Agriculture, and imitated by others, as their designation for a Cornish Rock broiler.

Rock Red \ *See: Red Rock*

rodents \ Rats and mice are common inhabitants in and around chicken coops, attracted by spilled feed. They invade any time of year but get worse during fall and winter, when they move indoors, gnawing holes in housing and burrowing underneath, thus providing entry for other predators. Rats eat eggs and chicks. Both rats and mice eat copious quantities of feed and by moving from one place to

Rhode Island
White hen

roost

another can spread diseases. To discourage rodents:

- Eliminate their hideouts, including piles of unused equipment and other scrap
- Store feed in containers with tight lids, and avoid or sweep up spills immediately
- Get a cat or a Jack Russell terrier
- Set traps baited with peanut butter or chocolate chips
- Use poison only as a last resort, because it works only if the rodents can find no other source of feed, which is unlikely in a backyard poultry situation, and because the poison might be eaten by a pet, a child, or harmless wildlife
- Don't waste money on techie solutions such as ultrasound black boxes and electromagnetic radiation, which are as ineffective as they are expensive

rolling mating \ A basic, two-generation linebreeding system in which the best pullets produced in a given year are bred to the best breeding cocks used to produce them and the best cockerels produced in the same year are bred to the best breeding hens that produced them. Rolling mating is the traditional method for breeding a small flock because it is simple and requires a minimum of record keeping.

roo \ Short for rooster.

MAKING A ROOST

The roost may be an old ladder or anything else strong enough to hold chickens and rough enough for them to grip, but without being so splintery it injures their feet. New lumber should have the corners rounded off so the chickens can more easily wrap their toes around it. Dowels or pipes of any sort (metal or plastic) do not make good roosts; they're too smooth for chickens to grasp firmly. Given a choice, chickens prefer to roost on something flat.

The perch for regular-size chickens should be constructed from a 2Q2 (5 cm by 5 cm) and for bantams from no less than a 1Q2 (2.5 cm by 5 cm). Allow 8 inches (20 cm) of perching space for each chicken, 10 inches (25 cm) for the larger breeds.

Make the perch readily removable so the droppings below can be cleared away easily, but fasten it securely so it, and your sleeping chickens, won't come crashing down.

If one perch doesn't offer enough roosting space for your number of birds, install additional roosts. Place them 2 feet (0.6 m) above the floor, at least 18 inches (45 cm) from the nearest parallel wall, and 18 inches apart. The roosts may be all the same height, but if floor space is limited, arranging them in stair-step or ladder fashion saves space. In that case space them 12 inches (30 cm) apart vertically and horizontally, so your chickens can easily hop from one level to the next.

rose comb,
nonspiked

roost \ The place where chickens congregate to spend the night. Wild chickens roost in trees. Domestic chickens are safer from predators when encouraged to roost indoors. \ To rest on a roost. When a low roost is available, light breeds generally start roosting at about four weeks of age, heavy breeds at about six weeks. Broilers should not be encouraged to roost, as these heavy birds will develop breast blisters and crooked keels and can incur leg injuries from jumping down off a roost. **See box on page 227 on making a roost.** *[Also called: perch]*

rooster \ *See: cock*

rooster, mean \ *See: aggressive chicken*

roosting bar \ A roost or perch.

roosting call \ A low-pitched, rapidly repeated sound made at nightfall when chickens are ready to roost. Its purpose is to encourage all the chickens to roost in the same place for safety's sake.

rose chafers \ *Macrodactylus subspinosus*, a beetle that emerges in late spring and early summer and feeds on the foliage and fruit of many plants and trees, including rose bushes, grapevines, and apple trees in eastern and central North America. Eating rose chafers can cause young chickens to go into convulsions and either die or recover within 24 hours.

R

rose comb,
spiked

Rosecomb rooster

rose comb \ A wide, flat, low-growing comb that may be slightly convex at the middle top and is covered with small, round bumps. The comb may follow the head's contour at the rear, as is characteristic of Wyandottes, or may terminate in a spike, as is characteristic of all other rose-comb breeds. Some breeds carry the spike horizontally; others have an upturned spike. The rose comb has been correlated genetically with low fertility.

Breeds bearing a rose comb include the Ancona, Bearded d'Anvers, Dominique, Dorking, Hamburg, Leghorn, Minorca, Nankin, Redcap, Rhode Island Red, Rhode Island White, Rosecomb, and Sebright. These breeds also have a single-comb variety: Ancona, Dorking, Leghorn, Minorca, Nankin, Rhode Island Red, Rhode Island White. See page 67 for illustration.

Rosecomb \ An old breed of true bantam originating in England as an ornamental chicken and named for its prominent rose comb. The Rosecomb bantam has white earlobes and a full tail that's large in relation to its small, compact body, and it comes in numerous color varieties.

Rosecomb eggs tend to be low in fertility, as is common among breeds bearing a rose comb. Chicks that do hatch are extremely delicate, requiring extra care to raise, but once they reach maturity these bantams are hardy and active and are capable fliers. The hens are poor layers of eggs with tinted shells and seldom brood.

Rosecomb hen

Rose Comb Clean Leg (RCCL) \ One of the groupings into which the American Poultry Association organizes bantam breeds. This class includes Ancona, Bearded D'Anvers, Dominique, Dorking, Hamburg, Leghorn, Minorca, Redcap, Rhode Island Red, Rhode Island White, Rosecomb, Sebright, and Wyandotte. The American Bantam Association uses the same classification, which does not recognize Redcap but additionally includes Nankin.

rotation \ Following a particular sequence, as in rotating dewormers, housing, or ground. The purpose of rotating dewormers is to prevent the development of a resistant population of internal parasites by not continuously using a single type of product. The purpose of rotating housing is to prevent a concentration of pathogens and parasites;

R

for instance, by raising a replacement flock in separate housing from the current mature flock.

Rotating ground likewise prevents a concentration of pathogens and parasites in the soil, as well as providing fresh forage. Rotation is accomplished by using cross fences to divide the ground into smaller units and letting the chickens into only one unit at a time.

Yard rotation involves dividing the yard surrounding a stationary shelter into two or more separate areas, each with its own entrance from the shelter. While the chickens are let into one area, the previously occupied area is reconditioned and reseeded, becoming naturally sanitized thanks to sunshine and fresh plant growth. Size each area as if it were the only yard, allowing 8 to 10 square feet (0.75 to 1 sq m) per chicken.

Range rotation involves moving a portable shelter on pasture and is typically used for raising broilers during the warmer months of the year. As a management method for laying hens, range rotation requires a shelter that can withstand all-season weather, as well as sufficient land for rotating the shelter to prevent destruction of the surrounding vegetation. Exactly how often the shelter must be moved depends on how fast the vegetation is eaten or trampled, which is a function of the number of chickens involved and their size, how crowded they are, the climate, the season, the weather, the type of pasture they are on, and whether or not they are confined inside the shelter. Surrounding the shelter with a movable fence gives the chickens more room to move around and therefore lengthens the amount of time they may remain on one piece of ground.

A good rule of thumb is to pasture only as many chickens as you can rotate without revisiting the same ground within a given year. As a general guideline, figure at least 110 square feet (10 sq m) per bird (which, on an acreage basis, means you can pasture up to 100 hens on 0.25 acre [0.1 hectare]). Since broilers don't hang around long, you can grow the same number in a little less than half the space; plan on 45 square feet (4 sq m) per meat bird (or about 250 broilers on a quarter acre).

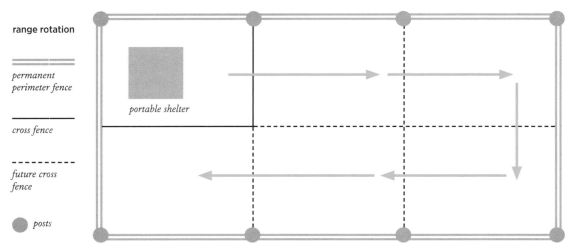

range rotation

permanent perimeter fence

cross fence

future cross fence

posts

portable shelter

roundworms \ Nematodes, of which several species afflict chickens and can do a great deal of damage. The most likely nematodes to infect a chicken are cecal worms, large roundworms (ascarids), capillary worms, and gapeworms.   *See also: specific type of worm*

roup \ *See: canker*

rump \ *See: uropygium*

Rumpless Game \ *See: Manx Rumpy*

rumplessness \ Lacking a tail-bone and therefore lacking a tail, as is characteristic of the Araucana and Manx Rumpy. Rumplessness reduces fertility. Some breeders find that over several generations the backs of rumpless chickens grow shorter until eventually natural breeding becomes impossible.

run \ *See: pen; yard*

rumplessness

R

S

the bird. A too-loose saddle, or one constructed of droopy material, will flop to one side, making it useless.

To prevent skin wounds, a saddle should be applied when feathers start disappearing. The chicken will try to back out of the saddle at first but will soon get used to it. Although a saddle is not intended as permanent clothing, as long as the chicken is with other chickens, the saddle must be left on until the bird's back has the protection of a full set of feathers.
[Also called: apron; mating saddle]

saddle \ The rear part of a cock's back where it joins the tail. \ An article of clothing fitted over the back of a chicken and designed to protect the feathers while they grow in. Saddles are used for hens that have been defeathered by overzealous roosters and for any chicken that has had blood feathers picked during a molt. They are available ready-made or may be homemade, in a variety of sizes that must properly fit the specific chicken. A saddle that is too tight will chafe, rub off breast feathers, injure the wings, or strangle

saddle feathers \ The long pointed feathers covering the saddle.

salmonella \ A genus of bacteria that includes many different species, only a few of which affect chickens. They are of particular significance because they can be deadly to chickens, and some also infect humans. The bacteria spread by means of contaminated droppings in litter, drinking water, and damp soil around drinkers. Survivors become carriers.

salmonellosis (plural: salmonelloses) \ One of several intestinal diseases, caused by various species and strains of salmonella, some of which exclusively affect chickens. Others also affect humans who eat undercooked eggs or meat from infected chickens.

salt \ Sodium chloride, a mineral that is vital to a chicken's health but in more than minute quantities can be toxic. Salt deficiency causes hens to lay fewer, smaller eggs and causes any chicken to become cannibalistic. Commercially

saddle

prepared rations contain all the salt a flock needs. Range-fed chickens that eat primarily plants and grain may need a salt supplement in the form of either kelp meal or loose trace-mineral salt.

Chickens offered a salt supplement must have access to water at all times. When water may not be available — because it is subject to freezing in cold weather or runs out on hot days — chickens without access to water may be poisoned by even a normal amount of salt. Poisoning may also result from ingesting rock salt used to deice sidewalks and driveways. *See also: electrolytes*

sanitation \ The enactment of measures to maintain a clean, healthful environment, including providing pure drinking water and disposing of manure. Good sanitation includes frequently cleaning and sanitizing feeders and drinkers, sanitizing any reused housing or equipment, and regularly cleaning the shelter and yard.

sanitizer \ A product or process that reduces the number of microorganisms on equipment and facilities. For a sanitizer to work, the equipment or shelter must first be thoroughly cleaned of manure and other debris. Sanitizing then may be accomplished in one of three ways, depending on what exactly is being sanitized.

HEAT: hot water, the hotter the better, or hot sun

RADIATION: ultraviolet rays from direct sunlight

DISINFECTANT: approved for use with poultry, such as Germex, Tek-Trol, Oxine, Vanodine, chlorine bleach (¼ cup bleach per gallon of hot water [15 mL/L]), or plain vinegar

sappy \ An undesirable yellowish or creamy tinge to white plumage, which could be hereditary or caused by diet. Or it could be the result of feather age, as compared to a clean, white, freshly emerged feather following a molt.

scalding \ The process of dunking a recently killed chicken, being butchered for meat, into steaming water to loosen feathers prior to picking. Scalding is optional for handpicking but required for machine picking. The dunking kettle must be large enough to accommodate the entire chicken without having the hot water overflow. The kettle may be heated on a gas stove, over a propane burner designed for camping, or on a grill over a wood fire. Turkey fryers are commonly used for scalding. Thermostatically controlled dipping vats are made specifically for scalding chickens.

To scald a recently killed chicken, hold it by the shanks and completely immerse it, head first, until all the feathered parts are under water. Move the bird up and down and back and forth so the water evenly penetrates the feathers. A squirt of dishwashing detergent in the water helps with penetration, especially for densely feathered breeds. Lifting the chicken out of the water and dipping it back in a time or two also improves penetration.

After about 30 seconds of dipping, pull a large tail or wing feather. Unless it slips right out, quickly dip the bird again before beginning to pluck.

Feathers should strip off easily and uniformly. If they don't, the scald wasn't long enough, the water didn't penetrate evenly, or the water wasn't hot enough. Torn skin and patches of feathers coming off with attached skin are signs the scald was too long or the water too hot.

scale of points \ *See: general scale of points*

scales \ Small, thin, hard plates that overlap to cover a chicken's shanks and the tops of its toes.

scaly leg \ An unhealthful condition of the shanks and toes caused by small parasites (scaly leg mites) that burrow under the scales, causing irritation and leaving deposits that raise the scales. The legs of an affected chicken appear to have thickened, and the discomfort causes the bird to walk stiff legged. Left untreated, scaly leg may result in permanent crippling.

Older chickens are more likely to be infected than younger ones and are difficult to treat because the mites burrow deeply. Soak the bird's legs in warm soapy water and gently scrape or rub off the softened dead scales. Then suffocate the mites by first dipping the legs in mineral oil or linseed oil; wipe off the excess, then coat the legs with petroleum jelly (Vaseline). Reapply petroleum jelly two or three times a week until the legs return to normal.

To prevent a reinfestation from untreated birds, treat all the chickens in the flock. Birds showing little or no sign of scaly leg may be treated weekly. *See also: scaly leg mite*

scaly leg mite \ *Knemidocoptes mutans*, a tiny, round, flat-bodied parasite that lives under the scales on the legs and feet of chickens and other poultry. Scaly leg mites spread slowly, primarily by direct contact with infected birds, and may be controlled by eliminating the infestation.

TEMPERATURE GUIDELINES FOR SCALDING

Young chickens with tender skin need a lower temperature and shorter scalding time than do older chickens with tough skin. However, scalding at a high temperature stiffens muscle tissue, making tough meat even tougher, and also causes the outer layer of skin to peel off.

A SEMISCALD, OR SLACK SCALD, at 125 to 130°F (52 to 54°C), keeps the outer layer of skin intact.

A SUBSCALD, at 138 to 140°F (59 to 60°C), loosens the outer skin layer but also loosens the feathers better, making them easier to pick.

A FULL SCALD, OR HARD SCALD, at 140 to 150°F (60 to 66°C), speeds things up, but the skin tears more easily, causing the meat to dry out.

S

scare wire \ An offset wire designed to zap a predator that gets too close to, or tries to climb over, a nonelectric fence. To be effective a scare wire should be approximately at the same height as the expected predator's nose, generally 10 to 12 inches (25 to 30 cm) above the ground. As insurance, especially in an area of deep winter snow, a second scare wire is often placed 8 to 10 inches (20 to 25 cm) below the top of the fence to discourage climbing.

scratch \ The habit chickens have of scraping their claws against the ground, stirring up the soil to bring up edible seeds, insects, worms, and bits of grit. Scratching also serves to wear down the toenails so they don't grow overly long. \ Any grain fed to chickens, commonly a mixture of at least two kinds of grain, one of which is usually cracked corn. It's called scratch because when it's scattered on the ground chickens scratch for it, perhaps thinking they'll turn up more. Scratch is high in energy and low in vitamins, minerals, and protein. Too much scratch in the diet radically reduces total protein intake. In growing birds, insufficient protein leads to feather picking. In laying hens, insufficient protein reduces egg production and the hatchability of incubated eggs and makes hens fat and unhealthy. Scratch should be fed only sparingly, if at all.

scratch

scratcher \ Affectionate word for a range-fed chicken.

Sebright \ An old breed of true bantam developed in England by Sir John Sebright and named after him. Sebrights are unique in being hen feathered. They have a rose comb and come in two laced varieties, golden and silver. Sebright eggs tend to be low in fertility, as is common among breeds bearing a rose comb. Chicks that do hatch are tiny and delicate, requiring extra care to raise, but once they reach maturity these bantams are hardy, active, and capable fliers. The hens are poor layers of eggs with white shells and seldom brood.

Sebright hen

Sebright rooster

secondary coverts \ Contour feathers covering the bases of the wings' secondary flight feathers. *See also: secondary feathers/secondaries*

secondary feathers/ secondaries \ Broad, stiff wing feathers, not quite as long as the primaries and growing nearest the chicken's body next to the primaries, between the wing's first and second joints. A chicken has between 14 and 18 secondaries, which remain visible when the wing is folded in a natural position.

secondary infection \ A disease that invades after immune defenses have been weakened by some other disease.

section \ The major parts of a chicken that are individually described in the standard for the breed and variety.

self color \ A single, uniform color throughout the plumage — such as black, blue, or white — in varieties that breed true.
[Also called: true color] *See also: blue*

self-limiting \ Any disease that runs its course in a specific amount of time, then stops without treatment.

semicardioid \ The semi-heart-shaped curve to the tail of American Game bantam cocks.

semicardioid

semioutcross \ To breed chickens that are distantly related. Compared to an outcross, a semioutcross results in some degree of hybrid vigor while minimizing the risk of introducing undesired genetic traits. \ The offspring of chickens that are distantly related.

septicemia \ Blood poisoning or invasion of the bloodstream by a microorganism, which occurs when any infection reaches the bloodstream. Signs include weakness, listlessness, lack of appetite, chills, fever, dark or purplish head, prostration, and death. Acute septicemia hits a bird so fast it literally drops in its tracks. Since most septicemic diseases cause reduced appetite and loss of weight before death, the classic indication of acute septicemia is the sudden death of an apparently healthy bird that has a full crop and is in good flesh.

Serama \ A breed of true bantam developed in Malaysia and named after the handsome and majestic Raja Sri Rama, a mythical character in Malaysian shadow puppet plays. Seramas are the smallest of all chicken breeds, weighing as little as 6 ounces (170 g) and standing only 6 inches (15 cm) tall. Their tiny size and calm, friendly demeanor make them ideal as house pets. Indeed, being small tropical birds they require a warm environment.

Seramas have a single comb, droopy wings, and a full tail held so high it sometimes touches the back of the head. They come in every color normal to chicken feathers but do not breed true to color — a cock and hen of similar color won't necessarily produce chicks of the same color.

The hens are year-round layers of tiny eggs ranging in shell color from white to deep brown. They will brood and make excellent mothers.
See page 238 for illustration.

SELF-SUFFICIENT BREEDS

Breeds that are aggressive foragers tend to be more low maintenance than other breeds. Two additional characteristics that enhance self-sufficiency are carefree reproduction and plumage color. Breeds that retain their instinct to brood require no human intervention to reproduce, in contrast to their specialized industrial cousins, whose brooding instinct has been taken away.

Some breeds, notably the Cornish, have been so distorted in the quest for a broad-breasted meat bird that the cocks have difficulty mounting a hen.

Feather colors other than white enhance self-sufficiency because they blend more easily into the surroundings, offering birds protection from predators while foraging.

GOOD FORAGERS

Barnevelder	Delaware	Leghorn	Plymouth Rock
Brahma	Dominique	Malay	Rhode Island Red
Buckeye	Dorking	Marans	Rhode Island White
Campine	Hamburg	Modern Game	Sussex
Catalana	Holland	Naked Neck	Wyandotte
Chantecler	Java	New Hampshire	
Cochin	Lakenvelder	Norwegian Jaerhorn	
Cubalaya	Langshan	Orpington	

AGGRESSIVE FORAGERS

American Game	Fayoumi	Minorca	Spanish
Ancona	Houdan	Old English Game	Spitzhauben
Andalusian	Kraienkoppe	Penedesenca	Welsumer
Empordanesa	La Fleche	Redcap	

serrated \ Having V-shaped, saw-toothlike points, a characteristic of a single comb.
See page 67 for illustration.

setting \ A group of hatching eggs in an incubator or under a hen.   *See also:*

clutch \ Keeping eggs warm so they will hatch; incorrectly called sitting.
See also: brooding

setting up \ Encouraging a chicken to strike a pose that best displays its unique breed characteristics. Setting up

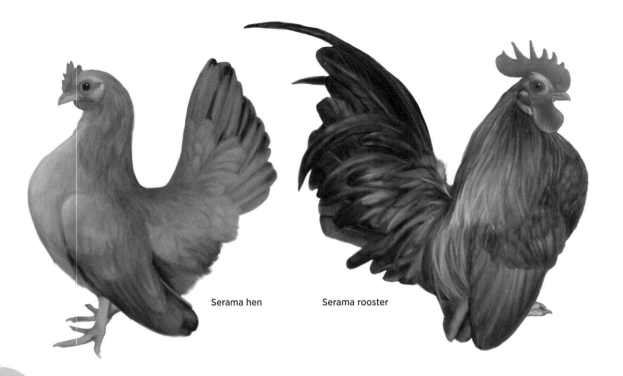

Serama hen   Serama rooster

is part of the training of a chicken destined for exhibition. Some breeds (such as Cochin) show best in a compact or horizontal stance, while others (such as Modern Game) show best when more upright or vertical.

A chicken is taught to set up by tapping it in appropriate places with a judge's stick, which is a telescoping pointer available from any office supply store. To get a bird to lower its tail, tap above the tail; to raise the tail, tap below the tail. Teach the bird to stretch upward or bend downward, as appropriate, by asking it to reach for a tasty tidbit — such as a tiny morsel of high-quality canned cat food — placed at the end of the judge's stick.

sex chromosomes \ The single pair of chromosomes that determine a chicken's gender, as well as any characteristics that are sex-linked, because they are transmitted by genes carried on the sex chromosomes. Opposite to humans, cocks have two similar sex chromosomes (designated as ZZ), while hens have dissimilar ones (ZW). The Z chromosome carries many times more genes than the W chromosome.

sex determination \ The gender of a chick hatching from an egg is determined by the hen, exactly opposite to gender determination in human babies. Each egg inherits a Z chromosome from the cock; half the eggs inherit a Z from the hen, the other half inherit a W. Approximately half the eggs therefore hatch into cockerels and half into pullets. The two most common reasons for significant deviations from this ratio are:

- Random death of embryos and chicks
- Sex-linked lethal genes

sexed \ Chicks that have been sorted into two groups, depending on whether they are pullets or cockerels. Chicks are sexed in one of three ways:

- Color sexing
- Feather sexing
- Vent sexing

When purchasing sexed chicks, expect to pay more for pullets than for straight run and less for cockerels. The latter have the least value because a flock needs fewer roosters than hens, or none at all, and a crowing cock may be unwelcome or illegal.

sex feather \ Any one of the pointed back, hackle, saddle, sickle, or wing bow feathers characteristic of a cock of most breeds — except those that are hen feathered — as distinct from the rounded feathers of a hen.

sex-link \ A characteristic that is genetically transmitted by genes carried on the sex chromosomes (ZZ for cocks, ZW for hens). Since the Z chromosomes carry nearly all the sex-linked genes, and a pullet gets her single Z chromosome from her sire, she inherits sex-linked traits only from the sire, while a cockerel gets a Z chromosome from both parents and therefore inherits sex-linked genes from both the sire and the dam. Some sex-linked characteristics are broodiness (transmitted through the sire from his mother to his daughters), albinism, dwarfism, nakedness, barring, late feathering, and the silver color pattern (white plumage with black hackles, wings, and tail feathers).

An example of sex linkage is mating a New Hampshire (or Rhode Island Red) cock to a Delaware hen, which results in pullets with solid red feathers and cockerels with the Delaware's dominant silver feather pattern. The opposite mating, of a Delaware cock to a New Hampshire (or Rhode Island Red) hen, results in all the first-generation offspring's having the Delaware pattern. \ A chicken that has been crossbred to take advantage of dominant and recessive, or the presence and absence of, sex-linked genes so its gender at hatch may be determined by physical appearance (color sexing or feather sexing), rather than by a microscopic examination of its sex organs (vent sexing).

sexual maturity \ The age at which a pullet starts laying or a cockerel becomes fertile. Most cockerels reach sexual maturity around six months of age, although early-maturing breeds may mature sooner, while late-maturing breeds may not be fertile until seven to eight months. In a cockerel, comb development is the best visible indication of sexual maturity.   *See also: age of lay*

shaft \ The stem of a feather, extending from the quill and running up the feather's center, to which the barbs are attached.

shafting \ The characteristic of a feather in which the shaft is either lighter or darker than the web. Shafting is undesirable in most cases — including in stippled varieties (such as light- and dark-brown Leghorn hens) and in solid buff and red varieties — but is a standard requirement in a few varieties, such as the quail Belgian d'Anvers.

Shamo \ An ancient breed originating in Thailand and developed in Japan for cockfighting but elsewhere considered ornamental. The breed name comes from the Japanese word *Sham*, or Siam, the former name of Thailand. The Shamo is the second tallest chicken after the Malay, growing to about 30 inches (75 cm) in height. It has a pea comb, although, occasionally, walnut combs appear; a beetle brow; an upright stance; and a downward-turning tail. Shamos may be large or bantam in size and come in a few established color varieties but are not bred for color outside exhibition circles. They have an athletic body with strong muscular legs and are extremely aggressive toward each other but not toward humans. The hens are poor layers of eggs with light-brown shells and will go broody but are awkward and clumsy setters.

shank \ The part of a chicken between the hock and the toes. The shank is technically part of the foot but commonly considered part of the leg. Round shanks are desirable in most breeds. Aseels are the exception. Their shanks are square, and the appearance of round or D-shaped shanks is considered to be a sign of crossbreeding.

shank cleaning \ Shank scales molt annually, just as feathers do, and may be removed after the bird's shanks have been thoroughly soaked in warm (not hot) water, which softens dirt and

Shamo hen

Shamo rooster

scales for easy removal. The dead, semi-transparent, brittle scales may be popped off with a nail file, a toothpick, or your thumbnail. Soapy water, applied with a soft toothbrush, may be used to remove any remaining dirt.

shank feathered \ Having feathers growing down the outsides of the shanks and on the outer toe or on both the outer and middle toes.
[Also called: feather legged]

sheen \ A soft luster or shine on the surface of plumage, which on black feathers often appears greenish or purplish.

shell \ The hard outer part of an intact egg. The shell accounts for about 12 percent of the weight of a large egg. Of the 25 hours needed for a typical hen to develop and lay an egg, about 20 hours is used to form the shell, which consists of three layers:

- The inner, or mammillary, layer enclosing the inner and outer membranes surrounding the albumen
- The spongy, or calcareous, layer made up of tiny calcite crystals — like tiny pencils standing on end, some of which fuse together, leaving pores between them that allow moisture and carbon dioxide to escape from the egg and oxygen to get inside to form the air cell
- The bloom, or cuticle, which gives the shell its color and also seals the pores to minimize evaporation from the egg and prevent bacteria from entering through the shell

shell, bloody \ Blood on a shell sometimes appears when a pullet starts laying before her body is ready, causing tissue to tear. Other reasons for bloody shells include excess protein in the lay ration and coccidiosis, a disease that causes intestinal bleeding. Cocci does not often infect mature birds, but if it does you'll likely find bloody droppings as well as bloody shells.

shell, chalky \ A chalky shell or a glassy shell occasionally appears due to a malfunction of the hen's shell-making process. Such an egg is less porous than a normal egg and likely will not hatch but is perfectly safe to eat.

shell, double \ An egg within an egg, or a double-shell egg, appears when an egg that is nearly ready to be laid reverses direction and gets a new layer of albumen, covered by a second shell. Sometimes the reversed egg joins up with the next egg, and the two are encased together within a new shell. Double-shell eggs are so rare no one knows precisely why or how they happen.

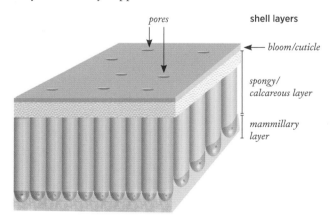

pores

shell layers

bloom/cuticle

spongy/ calcareous layer

mammillary layer

shell, misshapen \ An odd-shaped egg may be laid by an old hen or by a maturing pullet that has been vaccinated for a respiratory disease. Odd shapes may also result from a viral disease. Occasional variations in shape, sometimes seasonal, are normal. Since egg shape is inherited, expect to see family similarities. When sorting eggs for hatching, select only those that are of normal shape and size for your breed.

shell, missing \ *See: shell, soft*

shell, pale \ An older hen typically lays eggs with paler shells than those laid when she was younger. One explanation is that as the hen ages and her eggs get larger, the brown-pigmented bloom must spread over a larger surface area. A younger layer that produces eggs with paler-than-usual shells may be suffering from stress. Overcrowded nests, rough handling, loud noises, and anything that makes a hen nervous or fearful can cause her to either lay her egg prematurely, before the brown-bloom coating is completed, or retain the egg long enough to add an extra layer of shell on top of the bloom. Certain drugs, especially coccidiostats, can cause pale- or even white-shell eggs. Viral diseases that infect the reproductive system also cause hens to lay eggs with pale shells.

wrinkled shell

shell, soft \ An egg covered only by the inner and outer membrane. In reality such an egg does not have a soft shell but no shell at all. Eggs without shells occur when a hen's shell-forming mechanism malfunctions or for some reason one of her eggs is rushed through and laid prematurely. Stress induced by fright or excitement can cause a hen to expel an egg before the shell is finished. A nutritional deficiency, especially of vitamin D or calcium, can cause soft shells. A laying hen's calcium needs are increased by age and by warm weather (when hens eat less and therefore get less calcium from their diet). Appropriate nutritional boosters include a calcium supplement offered free choice and vitamin AD&E powder added to drinking water three times a week.

Soft shells that are laid when production peaks in spring, and the occasional soft or missing shell, are nothing to worry about. But if a hen lays fewer than her usual number of eggs, and the eggs she does lay have soft shells, she may have a serious viral disease.

shell, thin \ Eggs with thinner-than-usual shells generally appear during warm weather, when hens pant. Panting cools a bird by evaporating body water, which in turn reduces carbon dioxide in the body, upsetting the bird's pH balance and causing a reduction in calcium mobilization.

A pullet's first eggs may have thin shells, if the pullet isn't yet fully geared up for egg production. A hen that's getting on in age may lay eggs with thin shells, because the same (or a lesser) amount of shell material that once covered the smaller eggs she laid earlier in life must now cover larger eggs, stretching the shell into a thinner layer.

Thin shells also may be due to a hereditary defect or to an imbalanced ration, typically too little calcium or too much phosphorus. They also may result from some disease — most likely infectious bronchitis.

shell, wrinkled \ A hen may lay

a wrinkled egg if she has been handled roughly or if for some reason her ovary releases two yolks within a few hours of each other, causing them to move through the oviduct close together. The second egg will have a thin, wrinkled shell that's flat toward the pointed end. If it bumps against the first egg, the shell may crack and mend back together before the egg is laid, causing a wrinkle.

shell color \ Colored shells are the

result of pigments added during shell formation. Brown-egg layers produce eggs of varying shades ranging from barely tinted to nearly black, influenced by more than a dozen different genes that affect shell color. The darkest shells come from Barnevelders, Marans, Penedesencas, and Welsumers. Most of the pigment of a brown-shell egg is deposited in the bloom, the last layer added to the outside of an egg just before it is laid, leaving the inside of the shell pale or nearly white. Bloom dissolves when wet and easily rubs off when dry, which explains why cleaning a brown-shell egg removes some of the color.

The pigment of a blue-shell egg is spread throughout the shell, which is therefore just as blue on the inside as on the outside. Green shells come from crossing a blue-egg layer with a brown-egg layer, resulting in blue-shell eggs with a brown coating. The many different shades laid by so-called Easter Egg chickens result from blue shells coated with different shades of brown bloom.

All Asiatics and most Americans (except Holland) lay brown eggs. All Mediterraneans lay white eggs. Other classes are a mixed bag as to white, blue, brown, or tinted (cream, light tan, or pinkish).

As a general rule, hens with white earlobes lay eggs with white shells, and hens with red earlobes lay eggs with brown shells. Exceptions are Crevecoeur, Dorking, Redcap, and Sumatra, which have red earlobes but lay white-shell eggs; Araucana and Ameraucana, which have red earlobes but lay blue-shell eggs; and Penedesenca, which has white earlobes but among the darkest brown shells of any breed.

A common belief is that eggs with colored shells are more nutritious than those with white shells. Although eggs from backyard or pastured hens are likely to have brown or blue/green shells, in contrast to the white shells of commercially produced eggs, the shell color has nothing to do with an egg's nutritional value. The nutritional difference is not in the shell color but in how the eggs are produced. White-shell eggs produced by hens on pasture are more nutritious than eggs with colored shells laid by caged hens.

S

shell color

The pigment of brown eggs is deposited in the bloom.

The pigment of blue-shell eggs is spread throughout the shell.

shell gland \ An organ at the lower part of a hen's oviduct, where an egg obtains its shell. Of the typical 25 hours an egg spends traveling through the oviduct, it spends 20 hours in the shell gland. During the first 8 hours or so, the egg absorbs water and doubles in weight. During the remaining 12 hours, the egg is encased within an inner-shell membrane, an outer-shell membrane, a shell, and, finally, the cuticle.
[Also called: uterus]

shell quality \ The sum total of an eggshell's appearance, cleanliness, and strength. Appearance is important because the shell is the first thing you notice about an egg. Cleanliness is important because the shell is the egg's first defense against bacterial contamination; the cleaner the shell, the easier it can do its job. Strength influences an egg's ability to remain intact until a chick hatches from it or you are ready to use it in the kitchen.

shell strength \ The amount of pressure needed to break an egg by pressing at both ends. A shell's strength is influenced by the vitamins and minerals in a hen's diet, especially vitamin D, calcium, phosphorus, and manganese. Shell strength is also influenced by a hen's age — older hens lay larger eggs with thinner, weaker shells.

A shell gets strength from its shape as well as from its composition. An irregularly shaped shell, or one that is unusually round or elongated, is less strong than a well-shaped shell.

shoulder \ The outside of the wing at the point where the wing connects to the body.
See page 19 for illustration.

show breed \ *See: exhibition breed*

show coop \ A collapsible cage intended for the purpose of displaying a chicken on exhibition and also used to train a chicken prior to its being placed on exhibition. A typical show coop for bantams is 18 inches high and 18 inches (or a little less) square (45 cm high, 45 cm square). A typical coop for larger chickens is 27 inches high and 25 inches (or a little less) square (70 cm high, 65 cm square).

The floor is usually smooth, not wire, and at a show is often covered with shavings. When a show coop has more than one unit, to hold more than one chicken, a double partition between units prevents adjoining cocks from fighting and injuring each other. The door to each unit is on the front and generally slides upward from the bottom, so the chicken is put into and taken straight out of the coop without having to be lifted upward. Most show coops have built-in cup holders for water and feed.

show judge \ A person licensed by the American Bantam Association, the American Poultry Association, or both, who appraises chickens competing at a show according to the scale of points published in the licensing organization's standard. A specific number of points is allowed for each section (comb, tail, back, symmetry, weight, condition, and so forth), with deductions (or cuts) made

SHELL COLOR

WHITE SHELL

Bearded d'Anvers
Bearded d'Uccle
Blue Andalusian
Booted Bantam
Buckeye
Crevecoeur

Holland
Houdan
La Fleche
Leghorn
Minorca
Norwegian Jaerhon

Polish
Redcap
Sicilian Buttercup
Silkie
Spanish
Spitzhauben

Sultan
Vorwerk
Yokohama

TINTED SHELL

American Game
Ancona
Aseel
Campine
Catalana

Cubalaya
Dorking
Fayoumi
Hamburg
Kraienkoppe

Lakenvelder
Modern Game
Nankin
Old English
Phoenix

Rosecomb
Sebright
Shamo
Sumatra

BROWN SHELL

Australorp
Black sex-link
Brahma
Buckeye
Chantecler
Cochin
Cornish

Delaware
Dominique
Dutch
Faverolle
Japanese
Java
Jersey Giant

Langshan
Malay
Marans
Naked Neck (Turken)
New Hampshire
Orloff
Orpington

Plymouth Rock
Red sex-link
Rhode Island Red
Serama*
Sussex
Wyandotte

* Colors range from white to deep brown.

BLUE SHELL

Ameraucana
Araucana

DARK BROWN SHELL

Barnevelder
Empordanesa

Penedesenca
Welsumer

for such things as incorrect weight, missing tail feathers, off-color eyes, and other defects. Ultimately, judges base their decisions on two factors:

- How each bird compares to the ideal (or standard) for its breed and variety
- How each bird compares in type, condition, and training to the others competing in its class

See also: judging system

show quality (SQ) \ A chicken

that conforms closely enough to the standard description for its breed and variety to favorably compete with other chickens of similar quality in an exhibition. A show-quality chicken has the following characteristics:

- It has few or no defects
- It lacks disqualifications
- It is vigorous and free of disease

Sicilian
Buttercup hen

- It is mature and well developed
- It has the correct type for its breed
- Its weight is within the appropriate range for its breed
- It has a full complement of intact plumage
- It is of a uniform color appropriate to its variety

Sicilian Buttercup \ A breed of

possibly ancient origin that was developed on the Italian island of Sicily and named for the hen's golden color and the cuplike shape of the comb, which is circular with points around the outside, similar to a crown, tiara, or flower. The comb's large size makes this breed vulnerable to frostbite. The Buttercup is a small, active chicken that may be large or bantam and comes in a single color variety — the cock is red with a black tail; the hen is golden buff with black spangles. The hens are poor layers of white-shell eggs and seldom brood.

sickles \ The long, curved tail feath-

ers displayed by cocks of most breeds, except for those that are hen feathered, and which are particularly luxuriant in the longtail breeds. Sickles are of two types:

MAIN SICKLES. The two longest, most prominent feathers at the top of a cock's tail.

LESSER SICKLES. The remaining curved feathers hanging on both sides of the tail and covering the main tail feathers.

side sprig \ An undesirable appendage to a single comb, consisting of a clearly defined projection growing out of the side.

Silkie \ An ancient breed most likely originating in China (the breed is sometimes called Japanese Silkie, even though it probably arrived in Japan via China). The breed name comes from its soft, fur-like plumage, resulting from the inability of the feather barbs to lock. All that fluff makes these birds look bigger than they really are. Silkies can mature to a weight of 4 pounds (1.8 kg) or more, but in North America they are bred to be small and therefore are not true bantams.

A Silkie has a crest, walnut comb, turquoise earlobes, feathered legs, and five toes (some have six). The skin and shanks are black, and the face, comb, and wattles are nearly so. Silkies come in a few color varieties, of which white and black are the two most common, and may be either bearded or nonbearded.

They are friendly, docile chickens that can't fly (and many won't perch), and their plumage offers little protection from rain and extremes of temperature. The hens are decent layers of tinted eggs and are such excellent setters they are often kept solely for the incubation of the eggs of rare or nonbrooding fowl. *[Also called: Japanese Silkie]*

side sprig

main sickles

lesser sickles

Sumatra rooster

silkie feathered \ Having the semiplume feathers of a Silkie, which are structurally similar to the fluff of a normal feather. The shafts are unusually thin, and the barbs are unusually long and soft and cannot lock to form a web. As a result, silkie-feathered chickens appear to have fur instead of feathers, and their plumage offers less protection from wet and cold weather than the smooth, webbed feathers of most other chickens. *See also: wool feathered*

singe \ To burn the hairlike feathers off the skin of a plucked chicken by passing the bird over an open flame, such as a gas burner, a propane torch, or a purposely made singeing torch, which is similar to a handyman's propane torch. Commercial-strain Cornish-cross broilers

Silkie hen

don't need singeing because their hairs have been genetically removed.

single comb \ A flat comb arising from the top of a chicken's head, starting at the beak and extending back over the crown to end in a blade at the rear, topped with a series of sawtooth, upright points that are usually smaller at the front and back than in the center. A cock's comb stands straight up and is larger and thicker than a hen's. A hen's comb, depending on her breed, may be either erect or flopping over to one side.

The number of points a single comb has generally ranges from five to seven. For standardized breeds, the number of points is specified as part of the standard, which varies from country to country. In North America, the standardized breeds sporting single combs with five points are American Game Bantam, Ancona, Andalusian, Australorp, Barnevelder, Bearded d'Uccle, Booted Bantam, Campine, Cochin, Delaware, Dutch Bantam, Faverolle, Japanese, Java, Lakenvelder, Langshan, Leghorn, Modern Game, Naked Neck, Nankin, New Hampshire, Norwegian Jaerhon, Old English Game, Orpington, Phoenix, Plymouth Rock, Rhode Island Red, Rhode Island White, Serama, Spanish, Sussex, Vorwerk, and Welsumer. Single-comb breeds with six points include Catalana, Dorking, Fayoumi, Holland, Jersey Giant, and Minorca.

These single-comb breeds also have a rose-comb variety: Ancona, Dorking, Leghorn, Minorca, Nankin, Rhode Island Red, and Rhode Island White.

Single Comb Clean Leg (SCCL)
\ One of the groupings into which the American Poultry Association organizes bantam breeds. This class excludes game breeds and includes Ancona, Andalusian, Australorp, Campine, Catalana, Delaware, Dorking, Dutch, Frizzle, Holland, Japanese, Java, Jersey Giant, Lakenvelder, Lamona, Leghorn, Minorca, Naked Neck, New Hampshire, Orpington, Phoenix, Plymouth Rocks, Rhode Island Red, Spanish, and Sussex.

The American Bantam Association does not recognize Catalana, Holland, Java, Jersey Giant, or Lamona but additionally includes in this class Junglefowl, Nankin, Pyncheon, and Vorwerk.

singing
\ The musical sound made by a hen, usually consisting of rapidly repeated notes that sometimes are drawn out. Hens seem to sing when they are contented, perhaps for self-amusement, such as a person humming while doing dishes or singing while taking a shower.

sire family
\ The offspring of one cock mated to two or more hens, so all are full or half siblings.

skin color
\ All chickens within a given breed share the same skin color, which may be yellow, pink, white, or black. The skin color of a meat bird is purely a matter of preference. In general Europeans favor white-skin breeds, Asians like black-skin breeds, and Americans prefer yellow-skin breeds. The hybrids developed for meat production in North America have yellow skin. Since many people now remove the skin before cooking or serving chicken, preferences in skin color have become less important than they once were.

Skin color can vary with the chickens' diet, state of health, or age. Marigolds fed to hens to make their egg yolks a richer yellow will also deepen the color of their skin. A yellow-skin bird that isn't feeling well or for some other reason isn't eating well has paler skin than its flock mates. A young chicken with little fat may have bluish-looking skin. **See chart on page 250 for skin color of meat breeds.**

single comb

slip
\ A cockerel that has been incompletely caponized, such that both testicles have not been completely removed. Unlike the comb and wattles of a true capon, those of a slip will develop normally like the comb and wattles of an intact cock.

slipped wing
\ An undesirable but rare condition of a chicken's wing that has one or more twisted feathers; the primary feathers overlap in reverse order — over rather than under each other from outer to inner; or the entire section holding the primaries flops to the outside of the secondaries. As a result, the wing can't be folded close to the bird's body and held in the normal position. This condition may be genetic or may be caused by a dietary imbalance. *[Also called: angel wing; twisted wing]*

smut
\ Dark feathers that are uncharacteristic of the breed or variety, such as black feathers appearing on the body of a Rhode Island Red.

Society for the Preservation of Poultry Antiquities (SPPA) \ A national organization established in 1967 with the following goals:

- To protect and perpetuate rare breeds
- To encourage good breeding by sponsoring awards and shows
- To help members locate rare stock

soft feathered \ Technically, the plumage of any breed that is not hard feathered. More commonly, the term soft feathered is used to mean loose feathered. *See also: loose feathered/loosely feathered*

solid color \ A uniform plumage color that is unmixed with any other color or shade of color, mostly in reference to an all-black or all-white chicken. Solid color differs from self color in that the solid-color bird can carry recessive color genes that may appear in offspring, whereas a self-color bird does not.

SKIN COLOR OF MEAT BREEDS

YELLOW

Aseel	Cochin	Jersey Giant	Orloff
Barnevelder	Delaware	Langshan	Plymouth Rock
Brahma	Dominique	Malay	Rhode Island
Buckeye	Holland	Naked Neck	Welsumer
Chantecler	Java	New Hampshire	Wyandotte

WHITE

Ameraucana	Dorking	La Fleche	Penedesenca
Araucana	Faverolle	Marans	Redcap
Australorp	Houdan	Orpington	Sussex

PINKISH

Catalana

BLACK

Silkie
Sumatra

sour crop \ A fungal infection of the upper digestive tract usually appearing after a chicken has been treated with antibiotics or has been drinking water contaminated with droppings. As a result, feed moves so slowly through the crop that the crop becomes distended and the feed begins to rot and smell putrid.

In contrast to crop impaction, which requires surgery, a sour crop may be emptied by turning the chicken upside down (so the crop's contents fall out instead of flowing into the lungs) and gently massaging the crop for up to 30 seconds until the contents come out of the mouth. If additional massaging is needed, give the chicken a few minutes rest between each massage treatment. Apple cider vinegar added to the drinking water at the rate of 1 tablespoon (15 cc) per quart (15 mL/L) for two or three days will help restore the crop's proper pH balance to prevent the recurrence of sour crop.
[*Also called: thrush*]

space requirements \ How much space chickens need depends on their age, breed, and numbers; whether they are confined or have access to an outdoor run; and the climate. Young chickens need increasingly more space as they grow. Active breeds need more space than placid breeds. A few chickens need more space per bird than lots of chickens living together.

Consider this: Statistically, one light-breed hen needs at least 2 square feet (0.19 m); that's 1 foot by 2 feet (0.3 m by 0.6 m). Three hens therefore need 6 square feet (0.6 sq m); that's 2 feet by 3 feet (0.6 m by 0.9 m), which isn't much bigger than a cage, and gets even more crowded if the same space accommodates a feeder, drinker, and nest box. On the other hand, two dozen hens need at least 48 square feet (4.5 sq m); that's 6 feet by 8 feet (1.8 m by 2.5 m) — a pretty sizable coop that can easily accommodate a feeder, two drinkers, and four or five nest boxes.

Chickens that are confined entirely indoors, or that choose to remain indoors to get out of the wind or extreme heat or cold, need more space than chickens that enjoy free access to an outdoor run. Adequate space allows chickens to engage in normal avian behavior, minimizes stress, and helps prevent health and behavioral problems. As a general guideline, chickens should be allowed the following minimum living space:

HEAVY BREED: 3–4 square feet/bird; 1 square meter/3 birds

LIGHT BREED: 2–3 square feet/bird; 1 square meter/4 birds

BANTAM: 1–2 square feet/bird; 1 square meter/5 birds

spangled \ A feather pattern consisting of a V-shaped color mark at the tip of each feather that is either black against a white or golden bay background (as in silver or golden spangled Hamburg) or white with a black border separating the spangle from a bay background (as in speckled Sussex).
\ The Buttercup feather pattern consisting of a series of oval black marks arranged in somewhat V-shaped pairs running from each feather's shaft to the feather's edge.

spangled

Spanish hen

spiral band →

Spanish rooster

Spanish \ An ancient breed originating in Spain as a layer but now quite rare and considered primarily ornamental. The cock has long, white, overdeveloped earlobes and white skin around the eyes, making the face look like it's been covered with white paint. A large, bright-red single comb; long red wattles; and solid black plumage combine to make the white-faced black Spanish unmistakably distinctive. The hen looks similar, although her face isn't nearly as white and her comb flops to one side.

The Spanish is an active, noisy chicken, may be large or bantam in size, and has been developed in color varieties besides the most popular original black. The hens are good layers of eggs with chalk-white shells and seldom brood. *[Also called: clown chicken; clown-faced chicken; white-faced black Spanish]*

speckled \ The plumage pattern produced by the bicolored spangles of a speckled Sussex.

specs \ *See: blinders*

spent \ No longer laying well. As a hen ages, her egg production gradually slows down. At the point where the number of eggs a hen lays no longer justifies the cost of feeding her, she is considered spent (used up).

spike \ The rounded back part of a rose comb that tapers to a point. *[Also called: leader]* See page 66 for illustration.

\ A small bump that occasionally appears at the back of a Silkie's comb.

spiral band \ A plastic ring, similar to a key ring, placed on a chicken's leg for identification purposes. Spiral bands are the least expensive style of leg band but tend to break away easily and come in a limited number of colors, although they may be used in color-code combinations. They come in several sizes and are usually sold in lots of 50, all of one size and color. Spirals may be used alone or in combination with numbered leg bands.

Some uses for spiral bands are to identify a good broody hen, a chicken suspected of eating eggs, or a group of chickens of the same age or sharing the same parents. Exhibitors may band birds according to show wins, applying a blue spiral for first place, a red one for second place, and so forth (in Canada it's the other way around, red denoting a winner, blue for the runner-up).

Spitzhauben \ A breed originating in the Appenzell canton of Switzerland, where German is the main language. The breed sports unusual forward-pointing crest feathers, from which it derives its name; a *spitzhauben* is the lacy pointed bonnet traditionally worn by Swiss ladies of the region. This breed also has a V-comb, may be either large or bantam, and comes in three color varieties: black, golden spangled, and silver spangled. The hens lay white eggs and may brood. *[Also called: Appenzeller, Appenzeller Spitzhauben, German Spitzhauben, Spitz]* See also: *Appenzeller, Barthühner*

Spitzhauben hen

splashed \ A plumage pattern consisting of irregularly shaped patches of feathers of contrasting color or of more than one shade of the same color. \ An imperfect color arrangement on a feather appearing in an otherwise spangled or mottled variety, consisting of irregular patches of contrasting color.
See page 206 for illustration.

splayed leg \ A condition in a recently hatched chick whereby one or both legs slide out to the side so the chick cannot properly stand and walk. Splayed leg may occur during incubation, during the hatch, or after hatch. It may be caused by a too-high incubation or hatching temperature, but the most common cause is a hatcher or brooder floor that's too smooth for a chick to walk on, causing its legs to slip to the side. As a result the leg muscles can't develop, the chick cannot walk properly, and if the condition is not corrected, the chick cannot get to feed and water and will die. *[Also called: spraddled leg]*

splayed leg

FIXING SPLAYED LEG

A chick with splayed legs must be hobbled or have its legs mechanically brought together to train it to stand while its leg muscles strengthen sufficiently to let it stand and walk on its own. The hobble must be soft and flexible so it will not cut into the leg or bind, such as first-aid tape or a Band-Aid cut lengthwise to make it a good width to fit the chick's shank.

To avoid getting the tape or Band-Aid too tight and cutting off circulation to the foot, protect the legs with a bit of double-sided foam mounting tape wrapped almost all the way around except for the inner sides of each leg. Then cover the foam with the first-aid tape or a Band-Aid, completely covering the sticky side so it won't stick to the chick's down when the chick squats to rest. Keep an eye on the color of the feet to make sure the hobble doesn't cut off circulation.

At first the chick may have a hard time getting its legs properly positioned underneath itself. If the chick cannot stand with the hobble in place, reset the hobble to space the legs a little farther apart than they normally would be, and each day retape them a little closer together. During recovery the chick must have a nonslip surface to walk on.

Training a hobbled chick to walk may take several days and several reapplications of the hobble. The sooner a splay-legged chick is hobbled while its bones and muscles are still flexible, the more quickly it will learn to walk properly. To prevent splaying, use paper towels to line a smooth-bottom hatcher, as well as to line any brooder for the first few days after hatch.

split comb

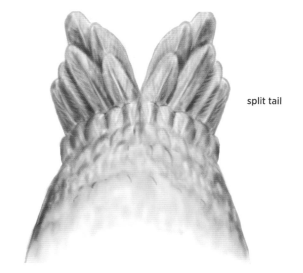
split tail

split wing

split comb \ An undesirable single comb that divides vertically at the rear of the blade.

split crest \ An undesirable crest in which the feathers divide and fall to both sides.

split tail \ An undesirable tail formation in which the failure of feathers to grow normally causes a gap between the main sickles, creating a permanent division down the center of the tail.

split wing \ An undesirable wing formation in which a permanently missing feather —as determined by the absence of a feather follicle — creates a gap between the primaries and the secondaries.

sport \ A genetic mutation that occurs naturally or is induced (for example, by radiation). \ A chicken exhibiting a genetic mutation, which generally breeds true. A sport sometimes becomes the basis for a new breed or variety. The Delaware, for example, was developed from a sport arising out of broilers bred by crossing Barred Plymouth Rock cocks with New Hampshire hens. The white Jersey Giant came about as a sport from the black Jersey Giant and the white Langshan as a sport from black Langshans. The white Plymouth Rock developed as a sport from the Barred Plymouth Rock. \ Cockfighting.

S

sporting fowl \ A line of chickens bred and raised for cockfighting.

spraddled leg \ *See: splayed leg*

sprig \ *See: side sprig*

sprouts \ Germinated seeds fed to chickens when green forage is not available. Sprouting improves the vitamin, mineral, and protein content of grains and provides an excellent source of green feed during winter and spring when naturally growing greens are not plentiful. If you sprout any or all of the grain portion of your chickens' diet, feed the sprouts separately. Combining sprouts with a dry ration can cause the dry feed to go moldy if it is not eaten right away. Chickens usually gobble down the sprouts first, so there's no need to worry that they will become moldy.

Any edible grain, seed, or legume may be sprouted. Use only fresh seed that has not been treated with a fungicide or any other chemical (as farm crop and garden seeds often are). Most scratch grains contain cracked corn, which of course won't sprout, but any whole grain intended for livestock should sprout quite nicely. Chickens particularly relish sprouted oats. Do a keyword search for "seed sprouting" on the Internet, and you will find all sorts of sprouting seeds and bulk sprouting trays and barrels, as well as instructions on how to sprout various kinds of grains and seeds.

spur \ The stiff projection at the back of a cock's shank, small and rounded in a young bird but sharply pointed and growing longer as the cock ages. The spur is an extension of the shank bone, covered with the same tough keratinous material that makes up claws and beaks. The spur starts out as a little bony bump. As the cock matures, it gets longer, curves, hardens, and develops a sharp pointed tip.

Spurs are used to gouge opponents during peck-order fights and to fend off predators. Most hens have little rudimentary knobs instead of spurs, although some have real spurs that can grow quite long and are considered undesirable, except on hens of the game breeds and on Sumatras.

Multiple spurs on either a cock or hen also are undesirable except on Sumatras. Mature Sumatra cocks have three to five spurs on each shank. The longest spur is flanked above and below by two shorter spurs, with one or two additional short ones below the top three. The female Sumatra has three to five vestigial spurs in the form of flattened bumps.

spur covers/spur protectors \ A pair of devices used to cover a breeding cock's spurs to prevent damage to himself or to hens, used to avoid trimming the spurs, as for example if the cock will be entered into a show. So-called breeder muffs, made of either leather or plastic, are sold by game fowl suppliers and are intended for use only during the time the breeder cock is in with hens. Sharply pointed spurs may need to be blunted so they won't poke holes in the muffs. In some states, use of breeder muffs is considered evidence of participation in cockfighting.

An alternative is to use thimble-like screw-on electrical wire connectors, sometimes sold as wire nuts, and twist one onto the end of each spur. Wire nuts come in different color-coded sizes. The most commonly used sizes are yellow and red, which between them should fit most full-size cocks with spurs long enough to hold a wire nut.

spurious wing \ Alula. It's called a spurious wing because it looks like a tiny fake wing growing out of the regular wing. *See also: alula*

spur trimming \ Clipping back or removing the hard outer cover of a mature cock's spurs, which like a human's fingernails require periodic trimming. Some spurs grow so long the cock has trouble walking. A cock with long spurs can stab his own body by jumping off the roost or a fencepost, leading to infection and death. Sometimes a spur curls back toward the leg and eventually pierces the shank, causing lameness. Long spurs can slice into a hen's back when the cock treads during mating. And a spur can cause serious damage to the cock's handlers. For all of these reasons, the spurs of a mature cock may need to be trimmed by blunting the spur or removing the entire outer shell. **See box on page 258 on how to trim spurs.**

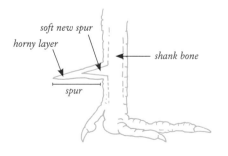

soft new spur
horny layer
shank bone
spur

squirrel tail \ A characteristic of Japanese Bantams but undesirable in most other breeds, whereby the forward part of the tail extends in front of an imaginary line drawn perpendicular to the bird's back.

stag \ A cockerel just entering sexual maturity, usually between the ages of 5 and 11 months, depending on the breed. When a cockerel enters the stag stage, his comb and spurs begin to mature. A cockerel raised for meat should be butchered before it becomes a stag, because the skin will become coarse and the meat will turn dark, tough, and strong tasting.

Standard \ The *American Standard of Perfection,* published by the American Poultry Association, or the *Bantam Standard,* published by the American Bantam Association. Both books are periodically revised to include newly accepted breeds and to remove older breeds to inactive status as they become rare or extinct. In addition to those listed in either *Standard,* other breeds may be found in North America, and many more exist worldwide.

standard \ The description of an ideal specimen for its breed. \ A chicken that conforms to the description of its breed in the *American Standard of Perfection*; sometimes erroneously used when referring to full-size as opposed to bantam breeds.

standard blue \ *See: blue*

SPUR TRIMMING

BLUNTING the tip of a mature spur may be done with a Dremel cutting wheel, wire cutters, or a pair of toenail or canine clippers, and the edges smoothed with a file. The Dremel cutting wheel is generally preferred, as clipping a spur may cause it to crack. Removing too much of a spur with any device will damage the quick, or soft tissue underneath, causing pain and bleeding. To estimate how far the quick extends from the shank, measure the diameter of the base of the spur, where it joins the shank, and multiply by three; for the average mature cock, the end of the quick is just over half an inch from the shank.

REMOVING the outer cover of a spur reveals the quick, which in two to four weeks will harden into a new spur. Exhibitors typically spruce up old show cocks by removing spur covers every couple of years. Grasp the spur

*removing the outer
cover of a spur*

near the shank with a pair of needle-nose pliers and gently, and patiently, wiggle the spur back and forth until it pops free, which takes about 60 seconds. Take care when removing the cover to not damage the tender tissue underneath.

Softening the spur first will help it work free more easily. Vegetable oil, liberally applied to the juncture between the spur case and the shank, helps soften the spur. Standing the cock in warm water is another way to soften the hard spur cover. The most popular method for softening the spur is to stick the spur into a hot baked potato — taking care to avoid burning the shank with the potato — and hold it there for about a minute. When the potato is removed, wiggle the spur until it slips free. Reheat the potato for the second spur.

A spur without the hard outer cover remains sensitive for a week or two, during which time the cock should be isolated from other chickens to avoid damage to the soft spur tissue. Keeping the cock in a separate, clean pen within the same area as the other chickens will minimize fighting when he is returned to the flock. If the freshly uncovered spur bleeds, stop the bleeding by applying a wound powder such as Wonder Dust or styptic powder or an astringent such as witch hazel, or hasten clotting with a little flour or cornstarch. The new spur will gradually harden and begin to grow long and eventually will need to be removed again.

standard bred \ Any chicken that is bred according to the published standard for its breed.

standardized \ A breed or variety for which a formal description, or standard, has been published and to which each individual in that breed or variety is expected to conform in appearance, with little variation from the standard. *See also: breed*

started pullet \ A young female chicken that is fully feathered and nearly old enough to lay. Buying started pullets rather than baby chicks has several advantages:

- They no longer need brooder heat.
- Feeding them for the short time they remain unproductive is less costly than feeding chicks to maturity.
- You don't have to wait as long to get eggs, while at the same time pullets have their full productive life ahead compared to mature layers.
- For exhibition they are less expensive than proven show birds but less likely to have serious faults than chicks, since birds with serious faults are culled early. Just be sure you do not acquire the culls.

starter \ A ration for newly hatched chicks. *See also: feeding chicks*

starve-out \ Failure of chicks to eat. Chicks don't have to eat the moment they hatch. For their first two days of life, they can survive on residual yolk. If, however, they don't eat within two to three days of hatch, they become too weak to actively seek food and will die of starvation.

Starve-out also may be caused by feeders placed where chicks can't find them or feeders set so high the chicks can't reach them. Other causes are placing feeders directly under the heat, and bedding newly hatched chicks on sand or sawdust that they fill up on instead of feed. When brooding on loose bedding, cover it with paper towels until all chicks are eating properly.

state birds \ A bird selected by a state's legislature as an emblem of that state. Two states have chickens as their state bird: Rhode Island has the Rhode Island Red and Delaware has the blue hen's chicken. In 2010, a movement was started in Georgia to replace the brown thrasher with the Cornish as that state's bird. *See also: blue hen's chicken*

station

station \ The ideal stance, symmetry, height, and reach characteristic of Aseels, Malays, Modern Games, and Shamos.

steal \ To lay eggs in a secluded and hard-to-find place — an instinctive habit of hens.

stern \ The part of a chicken's abdomen extending from the lower end of the sternum to the tips of the pubic bones.

sternal bursitis \ *See: breast blister*

sternum \ *See: breastbone*

sticking \ *See: debraining*

sticking knife \ A knife with a daggerlike blade, usually sharp on both edges, used for debraining chickens after slaughter. Any knife with a sharp, narrow blade may be used for this purpose.

sticky bottom \ *See: pasting*

stigma \ The part of the yolk sac within an egg that contains no blood vessels.
[Also called: suture line]

stippled feather

still-air incubator \ *See: gravity-flow incubator/gravity-ventilated incubator*

stippling \ Dots of contrasting color evenly scattered over a feather. Stippling is characteristic of several varieties, including light and brown Leghorn hens, and black breasted red Modern and Old English Game hens. *See also: mealiness*

straightbred \ The offspring of a cock and hen of the same breed and variety. Straightbred is a more accurate term than purebred, since chickens have no registry and therefore no registration papers as proof of lineage. *See also: purebred*

strawberry comb

straight run \ Pullets and cockerels that have not been sorted by gender but remain in the ratio in which they naturally hatched from a setting of eggs. Approximately 50 percent are pullets and 50 percent are cockerels, although some hatches naturally turn out to have more cockerels than pullets, while others are nearly all pullets.
[Also called: as hatched; unsexed]

strain \ A family of related chickens selectively bred by a single person or organization long enough for all the chickens to be uniform in appearance or production capability and clearly distinguishable from other strains of the same breed and variety. An established strain is usually identified by its developer's name, which is commonly a corporation for meat- or egg-production strains and an individual breeder for noncommercial strains, such as those intended for exhibition, fly tying, or cockfighting.
[Also called: line]
\ A naturally occurring subspecies of microorganism, such as bacteria or virus, having a distinct form or virulence compared to other strains within the same species. For example, the bird flu virus occurs in numerous strains having a wide range of virulence from low to high.

strawberry comb \ A low, compact comb with a rough surface that extends to the midpoint of the skull. Its shape resembles half a strawberry with the pointed end toward the rear. Strawberry comb is characteristic of the Malay.

stressor \ Anything that causes physical or mental tension, resulting in reduced resistance to disease and the possibility hens will lay fewer eggs or none at all. Chickens are nearly always under stress of one form or another, which may be minimized by a handler who moves quietly and gently among them and avoids making more than one major management change at a time.

COMMON STRESSORS

USUALLY MINOR

- Delay between hatching and first food or water
- Nutritional deficiency in chicks due to inadequate breeder-flock diet
- Cold, damp floor
- Debeaking
- Rough handling
- Low-grade infection
- Eating spoiled feed
- Unusual noises or other disturbances
- Extremely high egg production

MODERATE

- Chilling or overheating during first weeks of life
- Extremely rapid growth
- Chilling or overheating during a move
- Sudden exposure to cold
- Extreme variations in weather or temperature
- Unsanitary feeders, drinkers, or litter
- Internal or external parasites

- Insufficient ventilation or draftiness
- Vaccination
- Competition between sexes or individuals
- Medication (severity depends on drug used)

SERIOUS

- Overcrowding
- Nutritional imbalance
- Insufficient drinking water
- Combining chickens of various ages

SEVERE

- Suffocation caused by piling
- Lengthy periods without feed or water
- Inadequate number of feeders
- Poorly placed feeders causing starvation
- Onset of any disease

From: *Storey's Guide to Raising Chickens*, 3rd edition

---

stripe \ A pattern on individual hackle feathers seen on both cocks and hens, as well as on the saddle feathers of some particolored cocks, produced by the feather's web being of contrasting color to the outer edge. Examples of stripes may be seen in silver laced Wyandotte and colored Dorking cocks, which have black-striped saddle feathers, and in blue-breasted red Old English Game hens, which have blue-striped hackle feathers.

stub \ An undesirable short, downlike feather growing on the shank or toe of a clean-legged breed or variety.

stud mating \ The breeding practice of housing two or more cocks separately from hens, or in separate coops within the hens' housing, and briefly putting each hen individually into a cock's coop until she has been bred. Each hen is typically bred after laying every second or third egg, as determined by trapnesting. The advantage to stud

striped feather

mating is that it allows all the hens to be housed together while each is bred to a different cock.

sudden death syndrome \

A disease occurring primarily in intensively raised broilers, in which an apparently healthy chicken goes into convulsions, flips over onto its back, and dies within minutes. The cause is unknown, but it may be a metabolic condition affecting the heart.
[Also called: flip-over disease]

sugar \

An energy booster for newly hatched chicks. In the old days, before chick-booster blends became readily available, old-timers dissolved sugar in the drinking water of newly hatched chicks to give them extra energy. If the chicks were shipped by mail, or were just looking droopy and in need of a spurt of energy, ¼ to ½ cup sugar was stirred into each quart of drinking water (60–120 mL per 1 L) for the first day or until the sugar water was used up and replaced with fresh, clean water. Too much sugar can lead to pasting.

Sultan \

An old ornamental breed originating in Turkey, where it's called *Serai Tavuk*, meaning Sultan's palace fowl, because for decades it was kept exclusively in palace gardens. The Sultan is a smallish breed with a large crest, a V-comb, a beard and muffs, drooping wings, feathered legs, vulture hocks, and five toes. It may be large or bantam and originally had solid white plumage; other colors come from crossing Sultans with Polish. Sultans are docile and friendly, capable fliers, and less inclined to scratch than other breeds. The hens are poor layers of white-shell eggs and seldom brood.

Sultan hen

Sumatra hen

sultan's nose \ *See: uropygium*

Sumatra \ An ancient breed native to the island of Sumatra. This hardy, active, alert chicken is an extremely capable flier and high jumper. It has a pea comb, small wattles or none at all, a long sweeping tail, and multiple spurs (usually three or more on each shank). It may be large or bantam in size and comes in a few colors, of which solid black is by far the most popular, as the Sumatra's plumage is considered to be the most luminous of any black breed or variety. Even the face, shanks, and skin are black or nearly so. The hens are seasonally prolific layers of eggs with tinted shells, are good setters, and are especially good mothers. **See illustration of Sumatra rooster on page 247.**

[Also called: Black Sumatra]

supplement \ An addition to the normal diet offered to improve health, nutrition, or energy. Health boosters — usually consisting of a combination of vitamins and electrolytes — are sometimes added to the drinking water of chicks to help them cope with the transformation from embryo to chick and, if the chicks arrived by mail, to help them overcome shipping stress.

Supplements offered to laying hens include calcium, phosphorus, salt, kelp meal, and vitamins. These supplements are designed to help hens overcome the physical and nutritional stress of producing eggs. Grit is an additional supplement, offered to aid digestion of fibrous materials.

A vitamin/mineral supplement may be offered to show chickens to reduce the stresses of traveling and being on exhibit. Such a supplement should be given only before and after a show but never during a show, as the taste may cause a chicken in unfamiliar surroundings to stop drinking, thus increasing its stress level.

supplement hopper \ A small feeder in which a dry supplement, such as grit or kelp meal, is offered free choice. A simple and relatively inexpensive supplement hopper is a solid-bottom rabbit feeder attached to the coop wall; the solid bottom is important to note, as rabbit feeders also come with screened, or sifter, bottoms. If you have several hoppers for different supplements, use an indelible marker to write on the front of each hopper what kind of supplement it is for.

supplement hopper

surface color \ Plumage color that is visible when the chicken is relaxed enough to hold its feathers in their natural position, as compared to the undercolor. Surface color is the collective effect of the feathers' webs, since the fluff is typically covered and therefore not visible when the feathers are in normal position.

Sussex \ An old dual-purpose breed originating in England's county of Sussex, where this large-bodied chicken was raised primarily for meat. The Sussex is closely related to the Dorking; historical evidence suggests it was selectively bred from Dorkings with four toes. The Sussex has a single comb, may be large or bantam in size, and comes in several color varieties, of which speckled is the original. The hens are good layers of eggs with light-brown shells. The light and white varieties are the best layers, but they seldom brood, while the speckled varieties lay somewhat less well but are more likely to set, and they make good mothers.

suture line \ The part of the yolk sac within an egg that contains no blood vessels.
[*Also called: stigma*]

swarthy \ *See: dusky*

sword feathered \ Having the scimitar-shaped backward-curving sickles characteristic of Japanese and Serama cocks.

Sussex hen

syrinx \ The chicken's voice box. Unlike humans, whose voice box consists of vocal cords within the larynx at the top of the windpipe (the trachea), the chicken's voice box has no vocal cords and consists of the syrinx at the bottom of the windpipe, where the trachea splits at an upside-down Y-shape junction to create the bronchi that go into the two lungs.

All the sounds a chicken makes, including the cock's crow, require a cooperative effort among the tracheal muscles, syrinx, air sacs, and respiratory muscles. Contraction of muscles in the abdomen and thorax (the part of the body between the head and the abdomen) forces air from the air sacs into the bronchi and syrinx, while tracheal muscles work to alter the syrinx's shape to create various sounds. Each individual chicken produces a unique set of sounds by which the bird may be identified sight unseen.

systemic \ Involving the entire body, such as a systemic disease that affects the body as a whole or a systemic inhibitor used for parasite control.

systemic inhibitor \ A drug that permeates a chicken's entire body, making all its body tissues unappealing to internal and external parasites. Ivermectin (trade name Ivomec) is a systemic inhibitor sold as an over-the-counter cattle dewormer; it is not approved for poultry but is widely used anyway. Overuse of a systemic inhibitor can result in resistant parasites, and an excessive dose is toxic to chickens.

S

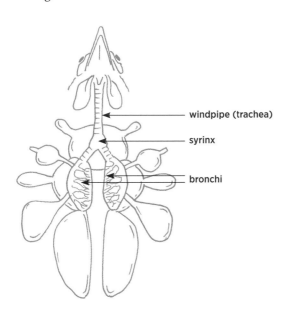

windpipe (trachea)

syrinx

bronchi

table egg \ An egg intended for use as human food, in contrast to a hatching egg that will be incubated to produce chicks. Table eggs from the supermarket are infertile and therefore will not hatch. Even if they were fertile, they've been refrigerated, which interferes with an egg's ability to hatch.

tailbone \ The last vertebrae at the end of the spinal column, which are fused together to form the main structural component of the uropygium. Unlike the tailbone in most animals, a bird's tailbone turns upward and controls movement of the tail and tail feathers to facilitate breeding, egg laying, pooping, and flying.
[Also called: pygostyle]

tail coverts \ The curved, pointed feathers covering the front and sides of the main tail feathers in a cock; the oval-shaped majority of feathers making up the tail of a hen.

tail feathers \ Collectively, the main tail feathers and sickles of a cock, or the main tail feathers of a hen.

tapeworm \ A long, white, ribbon-like segmented flatworm (cestode) that invades the intestine, causing weakness, slow growth, or weight loss and sometimes death. Chickens become infected by eating intermediate hosts, including ants, beetles, earthworms, grasshoppers, houseflies, slugs, snails, and termites. Once chickens become infected, removing tapeworms is difficult, although some of the dewormers that are approved for use with poultry can help control a tapeworm infection.

tassel \ A small crestlike tuft of feathers growing behind the comb and flowing gracefully down the back of the neck, as is characteristic of the Pyncheon. Unlike a crest, which grows to both sides of the head and may cover the comb, the tassel only grows backward and therefore leaves the comb fully visible.

telescope comb \ An undesirable comb formation consisting of either an indentation at the back of a pea comb or an inverted spike at the back of a rose comb.

temperature \ *See: body temperature*

temperature control \ The temperature inside a chicken shelter varies throughout the day and with the seasons. A chicken's body operates most efficiently at an effective ambient temperature between 70 and 75°F (21 and 24°C).

In colder weather, chickens eat more to obtain the additional energy

TABLE SCRAPS

Kitchen leftovers and garden refuse can be fed to chickens to reduce feed costs and increase variety in their diet. Just about anything you eat is suitable for feeding your chickens, with a few important caveats:

- Avoid feeding fried foods, which are difficult to digest and unhealthy
- Never feed dried beans or raw potato peels, which chickens can't digest easily — cook dried beans and potato peels or avoid them
- Don't feed strong-tasting foods like onions, garlic, or fish, which can impart an unpleasant flavor to poultry meat and eggs
- Don't feed avocados or guacamole unless you are absolutely certain none

of the brown seed cover is included, which can be deadly to chickens
- Never feed anything that contains caffeine or alcohol
- Avoid feeding anything that is high in fat, sugar, or sugar substitute
- Do not feed raw eggs or whole egg-shells, which can lead to egg eating; cook and mash eggs and finely crush the shells
- Never feed anything spoiled or rotten, which can make chickens sick
- Don't overdo any single item
- Keep scraps to no more than 10 percent of your chickens' diet, or the resulting nutritional imbalance can lead to slow growth, reduced laying, and poor health

---

they need to stay warm. To help them keep warm, insulate the roof and walls, bank the north wall with straw bales, let windows on the south wall furnish solar heat, check for and eliminate drafts while maintaining good ventilation to remove excess moisture. In most cases, installing a heater is unnecessary and unhealthful.

Hot weather is more problematic. For each degree increase, broilers eat 1 percent less, causing a drop in average weight gain. Egg production may rise slightly, but eggs become smaller and have thinner shells. When the temperature exceeds 95°F (35°C), chickens may die. To keep the shelter cooler in summer, insulate the roof and walls, use roofing and siding of a light color, and add awnings to shade the south and west walls.

thigh \ The feathered part of a chicken's leg above the shank between the hock and the body, consisting of a lower thigh and an upper thigh.

throat \ The front part of a chicken's neck below the beak.

throwback \ A chicken displaying recessive traits that had been genetically hidden for several generations.

thrush \ *See: sour crop*

thumb \ *See: alula*

thumb mark \ An undesirable hollow or depression on the side of a single comb.

tassel

crooked toe

straighten with first-aid tape or a Band-Aid

ticking \ Random specks of color that differ from a feather's ground color. Black ticking on the tips of the lower neck feathers is characteristic of New Hampshire and Rhode Island Red hens but is undesirable in most other breeds and varieties. However, a breeding chicken of a white variety with occasional dark gray or black ticking on feathers other than the primaries or the main tail feathers typically produces exceptionally white offspring.

tidbitting \ Repeatedly picking up and dropping a bit of food while sounding the food call. A mother hen tidbits to break up a large food item into pieces small enough for her chicks to peck. A cock tidbits to draw the hens' attention to some tasty bit of food and sometimes mock tidbits as a ploy to attract a hen. In the latter case when a hen comes running to see what he's found, he'll switch tactics and begin his courtship dance.

tight feathered/tightly feathered \ *See: close feathered*

tin hen \ Nickname for a metal circular still-air incubator.

tipped \ A color pattern of individual feathers, consisting of a white mark at the end of each feather, as is characteristic of mottled and spangled varieties.

toe \ A single digit of a chicken's foot. All chickens within a given breed share the same number of toes. Most breeds have four toes, with the third being longest. Dorkings, Faverolles, Houdans, Silkies, and Sultans have five toes. The extra toe grows above the hind toe and curves upward, so in most cases it never touches the ground.

toe, crooked \ A toe that turns to one side. A crooked toe in a newly hatched chick may be hereditary but more often is caused by a too-low incubation or hatching temperature. It may also result from having too much empty space in the hatching tray, thus giving newly hatched chicks too much room in which to move around before their bodies are strong enough to be active. Brooding conditions associated with crooked toes include overcrowding and a too-smooth brooder floor. In older chickens, crooked toes may develop because of nutritional deficiency or injury.

A crooked toe may be straightened in a newly hatched chick while the bones are still soft. Use first-aid tape or a Band-Aid cut to size and wrapped around the toes to hold them in normal position. A toe that is going to straighten out should do so within a day or two, as the chick's bones harden.

Unless the cause can be identified, a crooked-toe chicken should not be included in a breeder flock, as it may produce future crooked-toe generations. And a chicken with crooked toes is not suitable for show. On the other hand, although it may not be aesthetically appealing, a chicken with crooked toes will get along fine in all other respects.

toenail \ The curved, pointed horny nail at the end of each toe.
[Also called: claw]

toenail trimming \ Clipping a

chicken's toenails, which, like a human's toenails and fingernails, are made of keratin and continue to grow. When a chicken uses its claws to scratch the ground for food, the toenails may wear down naturally. But when a chicken is kept on soft bedding or in a cage without hard surfaces to scratch against, its claws will continue to grow until they curl and the chicken can't walk properly.

Nails that don't wear naturally need to be periodically trimmed, and chickens groomed for show likewise must have their nails neatly trimmed. Use a pair of canine toenail clippers or heavy shears. Trim away small amounts at a time to avoid snipping a nail too short. If you accidentally draw blood, stop the bleeding by applying an astringent such as witch hazel, styptic powder, or alum or encourage rapid clotting with a little flour or cornstarch.

Finish up by filing away any sharp edges. A mature cock with sharply pointed toenails also should have his nails filed, particularly the hind ones, to keep them from slicing into a hen's back during mating.

toe picking \ The cannibalistic

behavior of chicks that peck at their own or each other's toes. Pecking is instinctive behavior, and once chicks begin pecking they look for things to peck. If they can't easily find food items to peck, they'll pick toes, sometimes to the point of causing bleeding.

COMMON CAUSES OF TOE PICKING
- Running out of feed
- Not having enough feeder space for the number of chicks

- Chicks outgrowing the baby feeder
- Feeders placed so high as to be out of reach

ADDITIONAL CAUSES OF TOE PICKING
- A too-warm or too-crowded brooder
- A too-bright brooder light
- A starter ration that's too low in protein
- Simple boredom

To keep chicks active, encourage exploration by gradually moving feeders away from the heat source.

toe punch \ A device used to punch

holes into the webbing of a hatchling's feet for identification. The toe punch, available from nearly any poultry-supply catalog, is about the size of a pair of fingernail clippers and functions much like a paper hole punch.

toe punching \ A method of iden-

tifying newly hatched chicks by means of a coded pattern of holes in the skin webs between the front three toes on one or both feet. Punching the web causes brief but intense pain, and holes in the webbing can snag as the foot grows and the holes get bigger. Still, many pedigree breeders prefer toe punching as a permanent method of identification that is not readily visible without close examination.

As each chick is removed from its pedigree basket, it is held gently but securely with one foot extended to expose the web between the toes. The punch is carefully centered over a web and with one firm stroke a clean hole is made through the web. The punched-out bit of skin usually hangs up and must be

toe punch

T

pulled off to keep the web from growing back together.

The patterns formed by punching or not punching a web form 16 possible combinations, allowing for up to 16 different matings. Each pattern is assigned a number from 1 to 16, and all chicks from one mating are punched with the same pattern. The pattern for 16 is the least desirable, since no webs are punched — you'll always wonder if this chicken is really from the #16 mating or if it once had punched webs that later healed back together. Many breeders therefore reserve #16 for birds of unknown parentage.

TOE PUNCH CHART

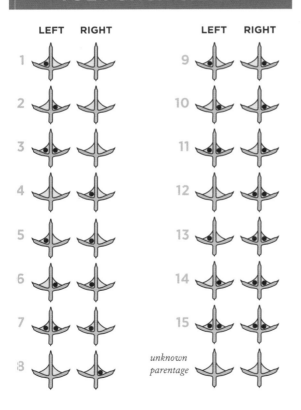

top color \ *See: surface color*

topknot \ *See: crest*

torticollis \ *See: wry neck*

toxic plants \ Landscape plants or weeds found in pasture that can be poisonous to chickens. Whether or not a specific plant is toxic may vary with its stage of maturity, growing conditions (such as drought), and other environmental factors. Should a chicken get a potentially toxic dose, the outcome will depend on the bird's age and state of health. Since chickens peck a little here and a little there to get variety in their diet, a bite or two of a toxic leaf or seed is unlikely to create a problem. Most toxic plants don't taste good anyway and therefore are not tempting except to a chicken that has few other food choices. A house chicken, for instance, may be tempted to eat toxic houseplants if no other greens are available. **See box on page 272 for common potentially toxic plants.**

trace minerals \ Dietary minerals a chicken needs in minimal amounts. Trace minerals are included in commercially prepared feeds, usually in sufficient amounts. For pastured chickens or a flock that otherwise does not derive most of its nutrition from commercial rations, two sources of trace minerals are trace mineral salt in loose form and kelp meal.

trachea \ Windpipe.

Transylvanian Naked Neck \ *See: Naked Neck*

trapnest \ A nest box to which a trap door has been fitted for the purpose of confining a hen after she has laid an egg, so you can precisely identify which hen laid the egg. Commercially made trap doors, or trapnest fronts, are available in heavy-gauge wire. Homemade fronts are usually made of lightweight wood, hinged at the top and held open by a hook or a prop the hen brushes against to cause the door to fall shut when she enters the nest.

Initially, hens may be suspicious about entering a nest that has been altered. To accustom them to the trapnests, install the fronts in advance and secure them in the open position so hens can come and go without getting trapped. When you do start trapnesting, check the nests often — every 20 or 30 minutes during prime laying time — as a hen that's confined for too long may soil or break her egg while trying to get out, and in warm weather she may suffer from being enclosed too long.

traumatic ventriculitis \ *See: hardware disease*

treading \ Short, quick movements a cock makes with his feet to keep him from sliding off a hen's back while he is attempting to mate. Over time, treading results in the loss of feathers from the hen's back. A hen with missing feathers has little protection during future matings and as a result may be seriously wounded by the cock's sharp claws or spurs.

To prevent injury to hens, keep the cock's toenails properly trimmed, taking care to round off the corners with a file. As a temporary measure, dress each hen — or at least those the cock favors — in a saddle.

Other methods of avoiding injury due to treading include removing the cock from the hens for all but a day or two a week or dividing the hens into two flocks and alternating the cock between the two groups. When using more than one cock in flock mating, move all the males as a group to prevent renewed peck-order fighting. Or house each cock in a separate coop and let each run with the hens for a few hours a week. Each cock will have a different set of favorite hens, offering the others some relief.

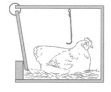

trapnest

trees \ Located inside a chicken yard, trees provide shade, and low-growing trees offer some protection from aerial predators. However, trees should be situated well away from the fence line; a tree close to the inside of the fence gives chickens a handy way to fly up into the branches and out of the yard. A tree outside the fence with overhanging branches gives climbing predators a handy way to get into the yard. Make sure any trees in or near the chicken yard are not close enough to the fence to allow branches to overhang.

Some chickens prefer to roost in a tree at night, leaving them open to predation by owls and other nighttime stalkers. For this reason most keepers train their chickens to go into a shelter at night. Leaving a light on inside the coop at nightfall, and feeding the chickens inside the shelter, is usually enough to encourage them to move indoors instead of roosting in trees.

trematode \ *See: fluke; worm*

COMMON POTENTIALLY TOXIC PLANTS

BLACK LOCUST

Robinia pseudoacacia

BLACK NIGHTSHADE

Solanum nigrum

BLADDERPOD, BAGPOD

Sesbania vesicaria

CASTOR BEAN

Ricinus communis

CORN COCKLE

Agrostemma githago

CROWN VETCH

Coronilla varia

DEATH CAMAS

Zigadenus spp.

JIMSONWEED, THORN APPLE

Datura stramonium

T

MILKWEED
Asclepias spp.

OLEANDER
Nerium oleander

POISON HEMLOCK
Conium maculatum

POKEBERRY
Phytolacca americana

POTATO
Solanum tuberosum

RATTLEBOX
Daubentonia punicea

VETCH
Vicia spp.

WATER HEMLOCK, COWBANE
Cicuta spp.

YEW
Taxus spp.

trichomoniasis \ *See: canker*

trio \ A cock and two hens or a cockerel and two pullets of the same breed and variety shown together as a group or housed together for mating purposes.

triple comb \ *See: pea comb*

trough feeder \ A long, narrow open container used for dispensing feed, as compared to a tube feeder. Since a trough is filled from the top and chickens eat from the top, a trough feeder tends to collect stale or wet feed at the bottom. Never add fresh feed on top of feed already in the trough. Instead, push the old feed to one side, and at least once a week completely empty and scrub the trough.

Chickens like to perch on a trough feeder. Roosting may be discouraged by fitting the trough with an anti-roosting device that swivels and dumps any chicken trying to stand on it or by mounting the trough on a wall, which leaves little room for roosting.

Chickens that eat from a trough waste a lot of feed by beaking out. Wastage may be minimized by adjusting the trough to the height of the chickens' backs. Some troughs have adjustable legs for that purpose. Still, approximately 30 percent of the feed will be wasted when the trough is filled to capacity, 10 percent when the trough is two-thirds full, 3 percent from a half-full trough, and 1 percent from a trough that's only one-third filled. Clearly, you can save money by never filling a trough more than two-thirds full, and you can save a lot of money by having more troughs and putting less feed into each one.

true bantam \ A small breed of chicken that lacks a larger counterpart. True bantams include American Game, Bearded d'Anvers, Bearded d'Uccle, Booted, Dutch, Japanese, Nankin, Pyncheon, Rosecomb, Sebright, and Serama.

true blue \ *See: blue*

true color \ A single, uniform color throughout the plumage — such as black, blue, or white — in varieties that breed true.
[Also called: self color]   See also: blue

true to type \ Having the ideal size and shape, or type, for the bird's breed.
[Also called: typy]

tube feeder \ A cylindrical container used for dispensing feed, as compared to a trough feeder. Since feed is poured into the top of a tube feeder, and chickens eat from the bottom, feed doesn't have a chance to go stale. A tube feeder works fine for pellets or crumbles, but mash tends to bridge if the tube is filled to more than two-thirds capacity.

trough feeder

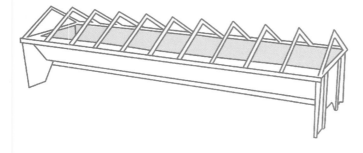

Chickens, especially young ones, like to roost on the edge of the feeder top or hop inside for a private snack, thereby fouling the feed. A cover solves that problem. Most feeders come without a cover, although some manufacturers offer covers as a separate option. These covers usually slope down from a pointed center, so any chicken that lands on one will slide off. The lid from a plastic bucket, of suitable size to fit the feeder, may be notched out to fit under the feeder handle. Chickens will still roost on top, but the cover will keep their droppings from falling into the feed.

To minimize feed wastage from beaking out, the feeder should be adjusted to the height of the chickens' backs. The easiest way to adjust a tube's height as the chickens grow is to hang the feeder from a chain. The chickens can then be discouraged from roosting on top of the feeder by tying something jiggly to the chain so it dangles into the top of the feeder and makes chickens nervous about landing there. Something jiggly may be made easily by cutting the bottom 2 inches (5 cm) from a well-rinsed empty plastic gallon jug (bleach, vinegar, or anything nontoxic), poking a hole through the center, and hanging it with a piece of strong string.

tucked up \ A characteristic of Modern Games whereby the stern tips straight upward, so the bird's head is directly above its feet. \ A characteristic of Cornish and the game breeds in which the primary and secondary wing feathers fold close together, the wing tips are held tightly against the bird's sides, and the feathers are short enough not to extend beyond the bird's body.

tufted \ Having ear tufts, a characteristic of Araucanas.

Turken \ *See: Naked Neck*

twisted comb \ An undesirable feature of a single or pea comb that is bent into an irregular shape.

twisted feather \ A feather in which the shaft rotates so the web does not lie smoothly in one plane. Twisted feathers are characteristic of frizzled varieties but are undesirable in other breeds, particularly if they occur among the main tail feathers, the sickles, or the primary or secondary wing feathers.

twisted neck \ *See: wry neck*

twisted wing \ *See: slipped wing*

type \ A chicken's outline or silhouette showing the bird's size and shape, by which its breed may be identified. Type influences the size and shape of the internal organs and the distribution of flesh, thus affecting the breed's suitability for the purpose for which it was developed, as well as its ability to adapt to various environments.
[*Also called: body type*]

typy \ Having the ideal size and shape, or type, for the bird's breed.
[*Also called: true to type*]

BODY TYPE BY BREED

Each breed may be identified by its type. Can
you identify these breeds?

The answers are on page 302, but no fair peeking until you've given it your best shot.

undercolor \ The color of the lower fluff of a chicken's feathers, collectively, as compared to the visible surface color. The undercolor doesn't show when the chicken is relaxed enough to hold its feathers in their natural position.

unsexed \ *See: straight run*

unthrifty \ Appearing to be unhealthy, failing to grow at a normal rate, lacking energy and vigor, or failing to lay at a normal rate for the breed.

upper thigh \ The part of a chicken's leg between the knee joint and the body.

urates \ *See: urine*

uric salts \ *See: urine*

urine \ A healthy chicken doesn't excrete much liquid urine but expels blood wastes in the form of semisolid urates consisting of uric acid combined with a concentration of minerals — called uric salts or urine salts. Urates appear as a white, pasty cap on top of a chicken's droppings. Occasionally a chicken will expel only the urates, which is perfectly normal.

The expelling of urates in a semidry form reduces the amount of water lost from the chicken's body. However, a certain amount of liquid urine is also excreted by the chicken's kidneys. The higher the moisture content of the bird's diet, as compared to a diet consisting solely of dry ration, the more liquid urine the kidneys will expel. Moments of stress can also cause a chicken to expel liquid urine.

urine salts \ *See: urine*

uropygial gland \ A pea-size gland at the base of a chicken's tail, on top of the uropygium, consisting of two symmetrical lobes that come together into a small nipple at the center. The uropygial gland secretes oil a chicken uses while preening its feathers, which it obtains by rubbing the nipple with its beak.
[Also called: oil gland; preen gland]

uropygium \ The spongy triangular bump at the rear end of a chicken's spinal column, consisting of flesh surrounding the tailbone and containing the uropygial gland. The tail feathers grow from the uropygium, except in rumpless breeds, which lack a uropygium.
[Also called: parson's nose; pope's nose; sultan's nose; rump]

U.S. gait scoring system \

A simplified modification of the British Kestin score, used to assess the welfare of broilers by rating the ability of individual birds to walk based on a three-point scale as follows:

1 No sign of lameness
2 The chicken limps but can awkwardly walk for a distance of 5 feet (1.5 m)
3 The chicken is so lame it will not walk as far as 5 feet (1.5 m)

uterus \ *See: shell gland*

utility breed \ A dual-purpose breed. \ A dressed chicken missing a part, such as a wing or a patch of skin, and therefore not aesthetically suited for roasting whole.

examples of utility (dual-purpose) breeds

Delaware

Ameraucana

Marans

U

V-comb

V-comb \ A V-shaped comb consisting of two distinct hornlike tapered projections arising from a single base just above the beak, as is characteristic of the Crevecoeur, Houdan, La Fleche, Polish, Spitzhauben, and Sultan.
[Also called: horn comb]

vaccination \ Inoculation with a vaccine to induce immunity against a specific disease. The most common diseases for which backyard chickens are vaccinated are all viral: fowl pox, infectious bronchitis, infectious laryngotracheitis, Marek's disease, Newcastle disease.

Vaccinate chickens only against diseases they have a reasonable risk of getting, including diseases chickens have experienced on your place in the past or new diseases that pose a serious threat in your area. Acquiring chickens from different sources and at different times, as well as taking chickens to shows and bringing them back home, increases their risk of getting a disease and spreading it to your other chickens. Some states therefore require exhibitors to vaccinate

against certain diseases; the premium list for each show should specify the local requirements. Otherwise, do not vaccinate against diseases that do not endanger your flock. Your veterinarian or state Extension poultry specialist can help you work out a vaccination program based on diseases prevalent in your area.

vaccine \ A product made from disease-causing organisms and used to stimulate immunity through inoculation. A vaccine confers immunity only if it is fresh, has been stored and handled properly, and is given exactly as directed on the label. The chickens also must be receptive to being immunized, which they will be if the weather is temperate and the chickens are properly fed and housed, have no internal parasites, and are otherwise in top health.

Vaccines usually come in enough doses for five hundred or a thousand chickens — much too many for a backyard flock. Save money by coordinating with neighboring chicken owners to share the vaccine and expense.

variety \ A subdivision of a standardized breed based on specific, identifiable features. Not all breeds have more than one variety. Examples of single-variety breeds include Ancona, Buckeye, and New Hampshire. Most breeds have several varieties, with more being developed all the time. These varieties may be distinguished based on one of four physical features:

PLUMAGE COLOR is by far the most common distinguishing feature. Some breeds have only two color varieties, such as

Campine (golden and silver), Norwegian Jaerhon (dark and light), and Holland (barred and white). Breeds with numerous color varieties include Leghorn, Japanese, and Wyandotte. The breed with the most color varieties is Old English Game.

COMB STYLE distinguishes variety in some breeds, including Ancona, Dorking, and Rhode Island Red, each of which has both single-comb and and rose-comb varieties. Breeds that have comb varieties in addition to numerous color varieties include Leghorn and Minorca, both of which have varieties in single comb as well as rose comb.

BEARDS AND MUFFS distinguish varieties of some ornamental breeds. Examples are Polish and Silkie, which come in bearded and nonbearded varieties. Some fanciers divide Booted Bantams into a bearded and a nonbearded variety; others consider the bearded ones to be a separate breed, the Bearded d'Uccle.

WEIGHT is a distinguishing feature of a few breeds, which are subdivided into classes according to maximum allowable weights. The Serama is divided into three weight classes (A, B, and C) and the Shamo into two (Chu and O).

vector \ A living thing, typically a biting insect, that carries disease organisms from one chicken to another. Examples of vectors include mosquitoes, ticks, and flies.

vehicle \ Anything that mechanically carries disease from one place to another. Examples of vehicles include dust, your clothing, and any poultry equipment.

velogenic Newcastle disease (VND) \ *See: Newcastle disease*

vent \ The outer opening of the cloaca, somewhat analogous to the anus.

vent feather \ Any feather growing between the vent and the tail or just below the vent. Vent feathers can interfere with fertility in heavily feathered breeds, because the feathers cover the vent, and in rumpless breeds, because they lack a tail to pull the feathers away from the vent during mating.

vent gleet \ *See: cloacitis*

ventilation \ The provision of fresh air. Compared to other animals, chickens have a high respiration rate, causing them to use up available oxygen quickly while at the same time releasing large amounts of carbon dioxide, heat, and moisture. Ventilation holes near the ceiling along the south and north walls of a chickens' shelter give warm, moist air a way to escape.

Vent covers let you open or close ventilation holes as the weather dictates. In warm weather, cross-ventilation keeps chickens cool and removes moisture. During the summer, open all the ventilation holes and open windows on the north and south walls. During cold weather, close all windows as well as the ventilation holes on the north side, keeping the holes on the south side open unless the weather turns bitter cold. Close all vents only as necessary to prevent a draft.

Use your nose and eyes to check for proper ventilation. If you smell ammonia

V

fumes and see thick cobwebs, the shelter is not adequately ventilated. *See also: draft*

vent picking \ Cannibalistic

behavior in which one or more chickens persistently peck at another bird's vent tissue. Continued vent picking eventually results in a large wound through which the picked bird's intestines and other organs may be pulled out, in which case it is called pickout.

Vent picking may start because feeders, waterers, or roosts are positioned where chickens below are attracted to pick at the vents of chickens above. It may also result when chickens are attracted to cloacal tissue that has become damaged and distended because a pullet has started laying at too young an age, is too fat, or lays an unusually large egg. *See also: blowout; pickout*

vent sexing \ A traditional Japanese

method of determining a hatchling's gender by examining minor differences in the tiny cloaca just inside a chick's vent. Accuracy requires a great deal of training, skill, and keen observation. *[Also called: cloacal sexing]*

ventriculus \ *See: gizzard*

vertebra (plural: vertebrae)

\ One of approximately 30 bones making up a chicken's backbone or spinal column.

vinegar \ A dilute solution of acetic

acid obtained by fermentation. Pure 5 percent vinegar has a pH of about 2.5. Disease-causing microbes favor a basic environment with a pH range of 7.5 to 9. Used at full strength, vinegar makes a good sanitizer for cleaning feeders, waterers, and other equipment.

Furthermore, most chickens like the taste of vinegar, and it's good for them. The microflora naturally colonizing a chick's intestines prefer an acidic environment with a pH range of 5.5 to 7.0. When a chicken is under stress — such as while recovering from an illness or injury — the bird may be disinclined to drink, and the gut flora may get out of balance. Adding apple cider vinegar to drinking water at the rate of 1 tablespoon per gallon (15 mL per 4 L) — double the dose if the water is alkaline — makes the drinking water more appealing to the chicken and encourages it to drink, while at the same time helping reduce the pH in the chicken's crop to encourage beneficial microflora to flourish there and ensure they make it to the gut.

Another beneficial use of vinegar is adding it to strange-tasting water (such as when traveling with a chicken or putting it on show) to make the water more palatable and encourage the chicken to continue drinking. Although chickens like the taste of vinegar, using more than the recommended amount can have an effect opposite to the desired one by causing them to stop drinking.

Avoid using vinegar in metal drinkers. Vinegar's acidity causes leaching of toxins from metal into the water, and also causes the metal drinker to corrode.

virulence \ The strength of a pathogen's ability to cause disease.

V

virus \ An ultramicroscopic organism that multiplies only in living cells. *Virus* is the Latin word for poison; most viruses are pathogenic.

viscera \ Internal body organs and glands.
[Also called: entrails]

vitamins \ Organic compounds that are essential for normal growth and body function but cannot be synthesized by the chicken's body and so are needed in small amounts in the diet. Vitamins are usually included in adequate amounts in a commercial ration but degrade if the ration is stale. If you question the freshness of the ration you use, either feed your chickens natural supplements or add vitamins to their drinking water. **See box on page 284 on vitamin supplements.**
See also: riboflavin

vitelline membrane \ The clear layer encasing an egg's yolk, separating it from the albumen.
[Also called: ovular membrane; yolk membrane]

Vorwerk \ A rare breed developed in Germany by Oskar Vorwerk as an attractive dual-purpose chicken. The Vorwerk has a single comb and comes in one color pattern — golden buff with a black neck and tail; essentially, a golden Lakenvelder. Bantam versions have been developed independently in the United States and in Europe. The only Vorwerks available in North America are bantams developed in the United States. Vorwerks are alert, active, capable fliers, and known for good feed-conversion efficiency. The hens are relatively poor layers of eggs with tinted shells and may go broody.

Vorwerk hen

vulture hock \ A group of long, stiff, straight feathers growing from the lower part of the lower thigh and extending backward and downward. Vulture hocks are characteristic of Bearded d'Uccles, Booted Bantams, and Sultans but are undesirable in other feather-legged breeds.

vulture hock

VITAMIN SUPPLEMENTS

Vitamin supplements for chickens contain a variety of vitamins including:

VITAMIN A. Essential for good hatchability and chick viability. It comes from green feeds, yellow corn, and cod liver oil. If you use cod liver oil, keep it fresh by mixing it into rations at each feeding.

VITAMIN B$_2$ (RIBOFLAVIN). Often deficient in stale rations. Good sources are leafy greens, legumes, and tomatoes.

VITAMIN D. Related to the assimilation of calcium and phosphorus needed for egg production. Deficiency causes shells to become thin. Since an embryo takes calcium from the shell, thin-shelled eggs that are hatched may produce stunted chicks. Two signs of deficiency are a peak in embryo deaths during the 19th day of incubation and chicks with rickets. Vitamin D comes from cod liver oil and sunlight. Chickens with adequate access to the outdoors should not have a problem with this vitamin.

VITAMIN E. Affects both fertility and hatchability. It comes from wheat germ oil, whole grains, and many fresh greens.

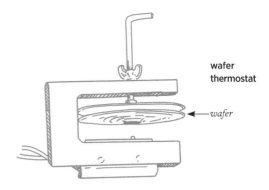

wafer
thermostat

wafer

wafer \ A metal disk filled with ether that is part of a thermostatic control used in some incubators and brooders. When the heat is on, the ether expands, causing the wafer to swell until it makes contact with a button switch that turns the heat off. As the temperature cools down, the ether contracts until the wafer loses contact with the switch, causing the heat to go back on. Regulating a wafer involves turning a control bolt to adjust the temperature at which the wafer makes contact with the switch.

Compared to an electronic thermostat, a wafer thermostat allows greater temperature fluctuations — as much as 2 degrees above or below the desired temperature. A wafer thermostat also typically takes longer to bring the temperature up if the heat is turned off and back on. And sooner or later the wafer will spring a leak, letting the ether escape and causing the switch to remain in the on position. If you use a wafer thermostat, always keep at least one spare wafer on hand.

walnut comb \ A roundish, lumpy comb that is greater in width than in length and somewhat resembles half a walnut. This style of comb is characteristic of the Kraienkoppe, Orloff, Silkie, and Yokohama.

warning call \ A sound made by a chicken, most commonly a cock, when it detects something unusual that doesn't appear to be dangerous. Different warning calls convey different meaning.

FLYING OBJECT ALERT is a sound a cock makes when he sees a high-flying bird or airplane overhead. He makes this sound while looking upward with one eye, and sometimes other chickens look up to see what he's looking at.

STARTLED NOTE is a short sound a cock makes when startled or surprised by a sudden sound or by something that suddenly appears within his range of vision. The intensity of the call varies, or the call may be repeated, depending on how startled he is. *See also: alarm call*

washing a chicken \ Giving a chicken a bath for the purpose of eliminating body parasites, cleaning dirty feathers, or grooming the bird prior to a show. **See box on page 286 for how to wash a chicken.**

walnut comb

WASHING A CHICKEN

Washing a chicken takes 15 or 20 minutes.

1. Begin with a basin full of warm water (90°F [32°C]). The temperature is right if you can hold your elbow in it without discomfort. Add enough pet shampoo or mild liquid dish soap to the water to whip up a good head of suds; don't use a harsh detergent, which makes feathers brittle.

2. Place one hand against each of the chicken's wings so it can't flap and slowly immerse the bird to its neck. If it struggles, dip its head briefly under water. Most birds relax as soon as they experience the soothing warmth of the water.

3. Thoroughly soak the bird by raising and lowering it and drawing it back and forth through the water, then use a sponge to soak the feathers through to the skin, working only in the direction the feathers grow. Rub in extra lather around the tail, where feathers tend to be stained by the oil gland, and around the vent. When the chicken is thoroughly clean, lift it from the bath and press out soapy water with your hands, working from head to tail.

4. With all plumage thoroughly washed, rinse the bird in fresh warm water that's slightly cooler than the wash water. Let the bird soak for a few minutes, until its feathers fan out or float, then move it back and forth in the water to work out remaining soap. Lift the chicken from the rinse, and press out excess water. To remove all vestiges of shampoo, rinse the bird once more, adding a little vinegar or lemon juice to this final rinse. When washing a white bird, you may wish to brighten its plumage by adding a couple of drops of bluing.

5. Remove the bird from the rinse water and squeeze out excess water from the feathers. Gently towel off the bird, then wrap it in a fresh towel and blot off any remaining water. Let the bird dry in a clean area away from any draft.

Cleaning the head of a crested bird. Hold it upside down by its legs, and dip the crest into the soapy water, keeping the bird's beak and eyes above water. Work suds into the topknot, or if the crest feathers are particularly dirty, apply a drop of shampoo directly to the topknot. Rinse the crest under running water to remove all traces of soap, taking care not to get any into the bird's eyes or nostrils. Washing a crest is easier if two people work together, one holding the bird while the other washes the head feathers.

washing eggs \ *See: egg washing*

water \ Newly hatched chicks can go without water for 48 hours, but the sooner they drink, the less stressed they will be and the better they'll grow. To make sure each chick knows where the water is, as you place each in the brooder, dip its beak into the drinker and watch that it swallows before releasing it.

As chicks grow, their need for water increases. To calculate how much water brooded chicks need, divide the chicks' age in weeks by four to determine approximately how many quarts (liters) they need per dozen chicks per day. For example, four-week-old chicks drink about 1 quart (1 L) of water per dozen per day.

On average, each mature chicken drinks between 1 and 2 cups (0.25 and 0.5 L) of water each day. A chicken drinks often throughout the day, sipping a little each time. Exactly how much a chicken drinks in the course of a day depends on several factors, including:

- **LAYING OR NOT** — layers drink twice as much as nonlayers
- **TEMPERATURE** — chickens drink two to four times more in hot weather
- **STATE OF HEALTH** — ailing chickens may drink excessively or not at all

Lack of water for even a few hours can cause hens to stop laying for days or weeks. Deprived of water for 24 hours, a hen may take another 24 hours to recover. Deprived of water for 36 hours, a hen may go into a molt, followed by a long period of poor laying. Water deprivation typically occurs in winter when the drinking water freezes and in summer when water needs increase but the supply does not. Chickens also may suffer water deprivation if the water quality is poor or they just don't like the taste — reason enough not to medicate water during hot weather.

waterbelly \ *See: ascites*

waterer \ A device used to furnish chickens with drinking water. A good waterer has these features:

- It's designed so the chickens can't roost over or step into the water
- It's easy to clean; one that's hard to clean won't be sanitized as often as it should be
- It doesn't leak, drip, or tip over easily
- It may be adjusted so the top edge is at the height of the chickens' backs
- It holds enough water that the chickens won't run out before you return to refill it

No less than once a week, scrub waterers with soap and water. Then sanitize them with either vinegar or a solution of chlorine bleach — one part bleach to nine parts water.
[Also called: drinker]

waterer, automatic \ A watering device that's connected to plumbing and refills automatically as chickens drink from it, designed to ensure the chickens never run out of water. Automatic devices are of two sorts: multiple single drinkers and community drinkers.

Multiple drinking stations are commonly used for caged chickens. These devices come in two basic styles, nipples and cups. Chickens must be taught to drink from a nipple, which can be a tedious undertaking. By contrast a cup holds a small amount of water and automatically refills each time a bird drinks.

Larger cups are available for community use in open housing. Another style of community drinker consists of a trough with a float valve to regulate water depth.

Features to look for in any automatic waterer are:

- **STURDINESS** — It shouldn't easily tip (dripping water) or break (releasing a flood of water) if a boisterous chicken bumps into it
- **ANTIROOSTING** — It should be designed to discourage roost; for instance, with a domed cap over a cup or a reel over a trough
- **EASE OF CLEANING** — If it can't be emptied and swabbed out easily, or quickly disconnected for a good scrub, it likely won't be cleaned often

bell waterer

waterer, bell \ A tubular or dome-shape water container with a water-filled rim encircling the bottom from which the chickens drink, giving the device the vague appearance of a bell. These waterers come in a variety of capacities ranging from 1 gallon (3.75 L) to 5 gallons (19 L) or more and are made of either plastic or galvanized metal. The plastic ones eventually crack, especially if left out in cold weather to freeze or in the sun for extended periods. The metal ones eventually rust through from constant ground-contact moisture. Unless designed for hanging, all bell waterers tend to get tipped over, and the rim typically fills up with debris.

Nonhanging bells may be placed on a platform, usually consisting of a wooden frame covered with hardware cloth and set on a bed of sand or gravel. With the waterer on top of the platform, spills fall through the wire into the sand or gravel to prevent puddling, and bedding and other debris can less easily get into the rim. A grout brush, designed for cleaning ceramic tile, is ideal for scrubbing out a messy rim.

waterer, chick \ A chick waterer is basically a scaled-down bell waterer, typically consisting of a glass or plastic jar that screws into a plastic or galvanized metal base. Don't be tempted to cut corners by furnishing water in an open pan or saucer. Chicks will walk in it, tracking litter and droppings that spread disease. They'll splash water and get wet and chilled, and some chicks may drown. A good chick waterer has these features:

- It is the correct size for the chickens' size and age — chicks should neither use up the available water quickly nor be able to tip the fount over

- The basin height may be adjusted as the chicks grow — a chick drinks more and spills less when the water level is between its eye and the height of its back

Some chick waterers are designed to be hung, providing an easy way to adjust the height. For a nonhanging waterer, a block or a scaled-down platform, such as one used for a bell waterer, may be used to raise the waterer as the chicks grow. Initially, place waterers directly on the brooder floor and no more than 24 inches (60 cm) from the heat source. As the chicks grow and their living area is expanded, make sure they never have to travel more than 10 feet (3 m) to get a drink. Whenever the waterer is upgraded to a larger size, leave the old one in place for a few days until the chicks get used to the new one.

Drowning is generally not an issue when chicks have a properly designed waterer, unless the brooder is so crowded some chicks fall asleep with their heads in water. For tiny bantam chicks that may slip into the water, place marbles or clean pebbles in the waterer's rim for the first few days until the chicks gain coordination.

waterer, gravity-flow \ A
watering system that uses an automatic drinker without hooking it up to a regular water line. The water is instead stored in a tank that is periodically refilled with a hose or with buckets or may be arranged to collect rainwater. A gravity-flow watering system is typically used where plumbing is not practical, such

chick waterer

as for pastured poultry. Details on how to set up a gravity-flow system may be obtained through many websites and Extension offices.

water heater \ An electrical warming device used to keep drinking water from freezing in cold weather. Several options are available, including:

- An immersion heater that is placed in a water trough
- A pan heater on top of which a metal bell waterer is placed
- A heating coil wrapped around pipes used for an automatic waterer
- A plug-in heated pet water bowl, which is a handy option for a small flock

To save money on electricity that would be lost by unnecessarily heating water on warmer days, plug the water-heating device into a Thermo Cube TC-3 thermostatically controlled outlet adapter.

water quality \ Water quality is a measure of how good the drinking water is in terms of meeting the chickens' needs. Factors affecting water quality include:

- **COLOR** — water should be colorless
- **ODOR** — water should be odorless
- **OPACITY** — water should be clear
- **BACTERIA** — water should be free of bacteria
- **MINERALS** — dissolved minerals exceeding 1,000 parts per million (ppm) can make the water taste unpleasant

Chickens should not have to drink from puddles or other stagnant unhealthful sources. Instead, provide fresh, clean water in suitable containers. Never expect your chickens to drink water you wouldn't drink yourself.

watering station \ The location of a waterer. A rule of thumb is to provide enough drinking stations so at least one-third of the chickens can drink at the same time, and position waterers no farther than 8 yards (7 m) apart. Even if you have so few chickens that one waterer would be adequate, furnish at least two to ensure that even the most timid chicken can get a drink without being chased away.

webbed feather

wattles \ Two flaps of flesh that dangle under a chicken's beak. The wattles are usually red, but some chickens have purplish wattles, including Silkie and Sumatra, as well as some varieties of American, Modern, and Old English game. A cock's wattles are usually much larger than a hen's. Most roosters get pleasure from having their wattles gently stroked, which causes them to close their eyes and relax.

One of the functions of wattles is to help keep a chicken cool in hot weather. Given water of sufficient depth, a hot chicken may splash some onto its comb and wattles to aid in cooling through evaporation. A chicken suffering from heat stress may be revived by applying cold water to its wattles.

Some breeds have rudimentary wattles or none at all. They include Aseel, Belgian Bearded d'Anvers, Belgian Bearded d'Uccle, Sumatra, and Yokohama.

wax picking \ A method of removing feathers from a recently butchered chicken. Wax picking follows rough handpicking as a fast way to get birds clean. It involves dunking the whole chicken in heated picking wax, then dunking it in cold water to harden the wax. Peel away the wax, and the feathers and pinfeathers alike come right off. When the wax is remelted and allowed to cool in the pot, feather debris settles to the bottom, and the clean wax on top may be saved for reuse. Picking wax may be melted in a kettle of steaming hot water; the paraffin floats at the top, so less wax is needed — allow 1 pound (0.5 kg) per four chickens.

webbed feather \ A smooth feather, such as the contour feather, wing feather, tail feather, or covert of most chickens, with the exception of Silkie-feathered breeds. *See: feather*

web foot \ An undesirable condition in which the flat skin between the toes extends more than halfway down the length of the toes.

web of wing \ *See: wing web*

well tucked \ *See: tucked up*

Welsumer \ A breed developed in the Netherlands and named after the village of Welsum; outside Holland the breed name is usually misspelled as Welsummer. These chickens have a single comb, may be large or bantam in size, and originally came in a single red-and-black (partridge) color pattern, although other color varieties have been developed. The partridge variety is auto-sexing — compared to cockerels, the pullets have darker markings on their heads and backs. The hens are fair layers of eggs with dark-brown shells; they seldom brood and when they do are not especially good mothers.

wet-bulb thermometer \ A thermometer used to measure humidity in an incubator. To make a wet-bulb thermometer, slip a piece of cotton, called a wick or sock, over the bulb or stem end of the thermometer, with the tail end of the wick hanging in water. Wicks may be purchased from incubator suppliers or made by cutting up a white cotton tennis shoelace. A wick must be replaced when it gets crusty with mineral solids and loses absorbency. Wicks last longer when used with distilled water, rather than tap water.

As water evaporates from the cotton, the thermometer gives a lower reading in wet-bulb degrees than it would without the wick (dry-bulb degrees). A typical wet-bulb reading is 86 to 88°F (30 to 31°C) during incubation and 88 to 91°F (31 to 33°C) during the hatch. Many incubator manuals refer to percent relative humidity. See chart on page 149 showing the relationship between wet-bulb degrees and percent relative humidity at various incubation temperatures.
[Also called: hygrometer]

whiskers \ *See: muff*

wattles

Welsumer hen

white-faced black Spanish \

See: Spanish

width of body \ The distance
between the two pubic bones, indicating
the amount of space available for an egg
to pass through when being laid.
[Also called: abdominal width]

wind egg \ A small yolkless egg.
Why it is sometimes called a wind egg
isn't clear, but since it is also called a
fart egg, perhaps people once believed
such an egg resulted from a hen breaking
wind. \ An egg laid without a shell. It's
called a wind egg because people once
believe such an egg was "fathered by
the wind." \ An addled egg. In a round-
about way the word addled supposedly
derives from the Greek term *ourion oion*,
meaning wind egg. \ An egg that is in
some way imperfect and thus unable to
hatch.   *See also: yolkless egg*

wing \ A feathered appendage cor-
responding to a human arm, a pair of
which enables a chicken to fly short

distances but is more commonly used for
balance during such activities as running
or jumping down from a roost.

wing badge \ A method of iden-
tification that is similar to a plastic ear
tag worn by four-legged livestock. The
wing badge attaches to a wing and has
numbers big enough to be read without
handling the chicken. Because they are
unsightly, wing badges are not often (if
ever) used for backyard flocks.

wing band \ A method of identifica-
tion consisting of an aluminum strip that
is attached through a chicken's wing web,
usually at the time the bird hatches. The
embossed numbers of a wing band offer a
permanent and reliable means of identification for the life of the bird, unless the
band is deliberately removed. One size
fits all; unlike a leg band, the wing band
needn't be changed as the chicken grows.

To avoid tearing the wing web,
apply bands after chicks have completely
fluffed out and have begun to toughen,
usually at a day old. Check week-old
chicks and adjust any band that might
have slipped over a tiny wing, and continue checking until the chicks are big
enough for the bands to stay in place and
not restrict wing development.

Wing bands may be used for both
large breeds and bantams, although most
show judges would frown on a small bantam wearing a visible wing band. Except
in tightly feathered breeds, a wing band
is otherwise clearly visible only on close
examination.

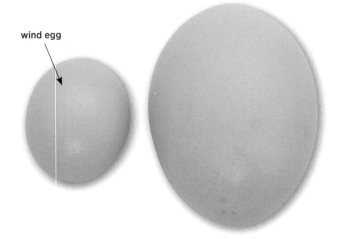

wind egg

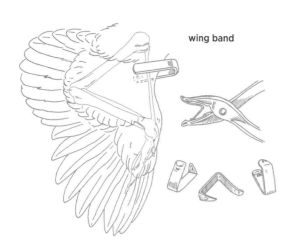

wing band

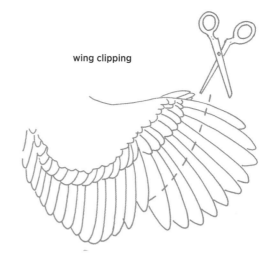

wing clipping

wing bar \ A distinct band of feathers across the middle of a wing created by the uniform arrangement of the tips of the secondary coverts.

wing bay \ The triangle of feathers appearing between the wing bar and the wing points created by those parts of the secondaries that are visible when the wing is held at rest.

wing bow \ A distinct group of feathers growing at the upper part of the wing, just above the wing bar. As these feathers cover the bases of the secondary wing coverts, they are otherwise known as lesser wing coverts.

wing clipping \ Using a pair of sharp shears to cut off the ends of the flight feathers of one wing to keep a chicken from flying out of its enclosure. Clipping the flight feathers causes the chicken to lack the balance needed for flight. It lasts only until new feathers grow during the next molt, which may be a few months in a young bird or up to one year for an older chicken. The

clipped feathers may not readily fall out during the molt, requiring your assistance before new feathers can grow in. Clipping, therefore, should be considered a last resort after all other methods of confinement have failed.

wing coverts \ Contour feathers covering the bases of the wings' flight feathers. Those covering the primary flight feathers are primary coverts; those covering the secondary flight feathers are secondary coverts.
See: feather types

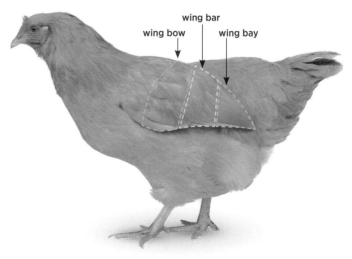

wing bow · wing bar · wing bay

wingette \ The outer two sections of the wing, separated from the drummette and prepared and served as finger food. *See: flat*

See page 44 for illustration.

wing feathers \ All the feathers, collectively, that cover the wing. The wing feathers are loosely grouped according to their size and placement into the following:
- Primary flight feathers
- Secondary flight feathers
- Primary coverts
- Secondary coverts
- Bow
- Shoulder

wing front \ The thin edge of the wing between the primary coverts and the shoulder.

wing molt \ The annual shedding and renewal of the wing's 10 primary feathers, which typically drop out at 2-week intervals, starting with the primary closest to the axial, and take approximately 6 weeks to regrow. A slow molter drops one feather at a time, requiring up to 24 weeks for the wing to fully refeather. A fast molter drops more than one feather at a time, and those that fall out as a group grow back as a group. A fast-molting hen quickly completes her molt and gets back to laying eggs.

wing points \ The tips of the wing's flight feathers.

wing web \ The triangular flap of skin between the chicken's shoulder and the wing's second main joint, which stretches and becomes more easily visible when the wing is fully extended.

wire mesh \ Any type of fencing material with openings small enough that chickens and most predators can't get through. Of the many kinds of wire mesh available, one that works well for chickens and is relatively low on the cost scale is yard-and-garden fence with 1-inch (2.5 cm) spaces toward the bottom and wider spaces toward the top. The small openings at the bottom keep poultry from slipping out and most small predators from getting in.

wishbone \ The chicken's collar bone, which connects the shoulders while allowing an opening for the digestive and respiratory tracts to pass from the head into the chicken's body. By tradition the slingshot-shaped wishbone, when

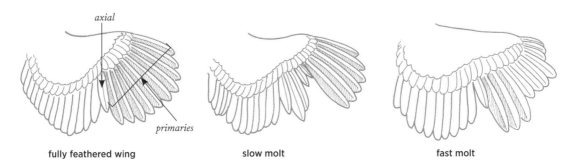

axial

primaries

fully feathered wing   slow molt   fast molt

removed from a roasted chicken, is pulled apart by two people who both make a wish; when the wishbone breaks, the person getting the larger piece expects to have his or her wish come true.
[Also called: pulley bone]

withdrawal period \ The minimum number of days that must pass from the time a chicken stops receiving a drug until the drug residue remaining in its body is reduced in its eggs or meat to an "acceptable level" to be safe for humans, as established by the United States Food and Drug Administration based on the drug manufacturer's recommendation. Withdrawal times for drugs that have been approved for use with chickens (and other food-producing animals) are published by the Food Animal Residue Avoidance and Depletion Program and may be found online at www.farad.org. If you cannot determine a specific drug's withdrawal period, allow at least 30 days.

wool feathered \ Silkie feathered. Early descriptions of Silkies sometimes referred to them as wool fowl or woolly hens, and Hedemora chickens are today described as wool feathered rather than silkie feathered.

worm \ A common name for various unrelated invertebrate animals with slender, long, soft bodies, of which two groups are of importance to chickens.

EARTHWORMS are annelids that burrow in soil, and feed on soil nutrients and decaying organic matter. When they come to the surface in a chicken yard, they are quickly gobbled up. Earthworms can provide chickens with an important source of animal protein, especially when they are specifically raised for the purpose. But earthworms, particularly those living in proximity to a chicken yard, can also serve as an intermediate host for a variety of internal parasites that affect chickens.

PARASITIC WORMS are a group of unrelated worms that live inside a chicken, receiving nourishment and protection while disrupting the chicken's nutrient absorption and thus reducing the bird's immunity to disease. The two categories of major significance to chickens are roundworms (nematodes) and flatworms (cestodes and trematodes). Parasitic worms are most likely to become a problem where chickens are kept on the same ground year after year. Under proper management, including good sanitation, chickens gradually develop resistance to parasitic worms. A chicken that has the opportunity to acquire resistance through gradual exposure to the parasites in its environment gets an unhealthy load only if it is seriously stressed; for instance, by crowding, unsanitary conditions, or the presence of some other disease.
[Also called: helminthes]

wishbone

wormer \ *See: dewormer*

worming \ *See: deworm*

wry neck \ A neurological condition in which a chicken's head and neck are so abnormally twisted that the bird loses coordination. Wry neck may be caused by any number of things, including

injury, infection (especially an ear infection), a toxic reaction (such as botulism), or a nutritional deficiency (such as vitamin E).
[Also called: crooked neck; torticollis; twisted neck]

wry tail

wry tail \ A tail that permanently leans to one side instead of being held vertically. Wry tail is more commonly seen in cocks than in hens but in either case is undesirable.

Wyandotte \ A dual-purpose breed developed in the United States in an area once occupied by the Native American Huron tribes, collectively known as Wendat, or Wyandot, which inspired the breed name. Wyandottes may be large or bantam and come in several color patterns, of which silver laced is the original. Their short tail, short back, and loose feathering give them the appearance of being somewhat round. They are the only breed having a rose comb that lacks a spike, and they sometimes throw single-comb chicks. The rose comb combined with loose feathering makes them cold hardy. The hens are good layers of eggs with brown shells, sometimes brood, and are good mothers.

Wyandotte hen

xanthophyll \ A yellow/orange pigment derived from plants of a large group known as carotenoids. It is called xanthophyll from the Greek words *xanthos*, meaning yellow, and *phyllon*, meaning leaf. Xanthophyll in a hen's diet is deposited in the yolks of her eggs, giving them their rich yellow color. The same pigment colors the skin and shanks of yellow-skin hens, as well as their fat. When a hen is laying, the xanthophyll goes directly to developing egg yolks, and she loses pigmentation in other parts of her body. In a commercial ration, xanthophyll comes primarily from lutein. Natural sources of lutein include marigold petals, green grass and other leafy greens, and yellow corn.   *See also: bleaching*

xanthous \ Resembling the color of an egg yolk, from the Greek word *xanthos*, meaning yellow.

X disease \ *See: aflatoxicosis*

yard \ A fenced-in outdoor area to which chickens have access during the day. A yard gives chickens a predator-safe place to get the sunshine, fresh air, and exercise they need to remain healthy. But chickens can quickly destroy the yard's ground cover by pecking at it, scratching it up, pulling it up, and covering it with droppings. The smaller the yard, the quicker it will turn to either hardpan or mud, depending on your climate.

Since chickens are most active near their shelter, denuding starts around the entrance and works progressively outward. In a large yard, vegetation may continue to grow in areas farthest from the doorway. Between the barren area and the grassy area, a band of weeds may develop that is so tough or unpalatable the chickens won't eat them and can't trample them. Mow both vegetated areas as needed to keep them tidy and safe.

Chickens will range farther if their yard is not entirely open but includes trees, shrubs, or small range shelters offering shade from the hot sun, relief from blowing winds, and protection from flying predators. Trees and shrubs are an attraction also because they may drop leaves and fruits, as well as harbor insects chickens like to snack on.

As a general guideline to yard size, a spacious area provides 8 to 10 square feet (0.75 to 1 sq m) per chicken. If the perimeter fence is 100 feet (30 m) or more from the shelter and the yard has no trees, station a basic range shelter about every 60 feet (18 m).
[Also called: run]   See also: pen

yarding \ Letting chickens into a fenced yard adjacent to a stationary shelter.

yearling \ A chicken in its first year beyond the year it was hatched.

Yokohama \ A longtail breed originating in Japan, developed in Germany, and named after the Japanese port city from which longtails were exported to other parts of the world. The Yokohama shares its history with the Phoenix and is similar in its small size and its conformation, differing primarily in having red earlobes and a walnut comb and being either solid white or red shouldered (also called red-saddled white) — consisting of a red body with a white neck and tail. Other color varieties have also been developed in both large and bantam versions. The cock's saddle feathers grow as long as 3 feet (almost a meter). To achieve such long growth requires a high-protein diet.

The hens are poor layers of eggs with tinted shells, although their rate of lay may be improved with extra protein. They may brood and are protective mothers.

yolk \ The yellow interior part of a chicken's egg, which contains all the egg's fat, a little more than half its protein,

Yokohama hen

and a number of minor nutrients. The function of the yolk is to nourish a developing embryo when the egg is incubated.

yolk color \ Egg yolks get their color from xanthophyll, a natural yellow-orange pigment in green plants and yellow corn. The exact color of a yolk depends on the source of the xantho-phyll. Alfalfa, for example, produces a yellowish yolk, while corn gives yolks a reddish-orange color.

Excessive amounts of certain pig-mented feeds can affect yolk color. Alfalfa meal, clover, kale, rape, rye pas-ture, and certain weeds, including mus-tard, pennycress, and shepherd's purse, make yolks darker. Too much cottonseed meal can cause yolks to be salmon, dark green, or nearly black.

The yolk is not of uniform color throughout but consists of concentric rings. At the center is a ball of white yolk, around which are alternating layers of thick dark yolk and thinner white yolk. Although you might never see it — except maybe in a hard-cooked egg — a neck of white yolk extends from the center to the edge of the yolk, flaring out and ending just beneath the blastodisc. **See chart on page 300 for influences on yolk color.**

yolkless egg \ An egg containing no yolk, most often occurring as a pullet's first effort, produced before her reproductive system is fully geared up. Sometimes a hen will lay a small yolkless egg at the begin-ning or end of her laying cycle. A yolkless egg may also occur if a bit of reproductive tissue breaks away, stimulating the egg-producing glands to treat it like a yolk and wrap it in albumen, membranes, and a shell as it travels through the oviduct. In place of a yolk, this egg will contain a small particle of grayish tissue.

[Also called: cock egg; dwarf egg; no-yolker; wind egg]

yolks, multiple \ *See: double yolker*

Yokohama rooster

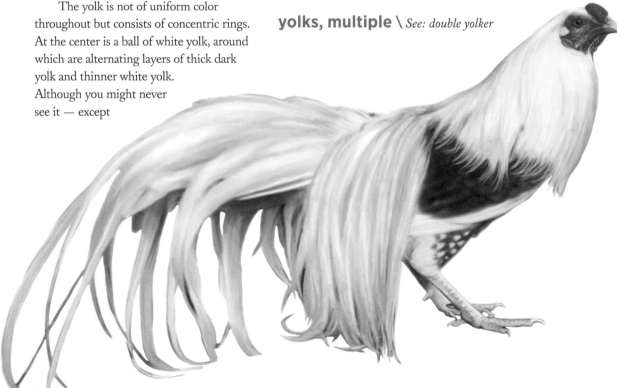

INFLUENCES ON YOLK COLOR

RAW EGGS

Color	Causes
Green	Acorns; shepherd's purse
Orange to dark yellow	Green feed; yellow corn
Reddish, olive green, black-green	Grass; cottonseed meal; silage
Yellow, dark	Alfalfa meal; marigold petals
Yellow, medium	Yellow corn
Yellow, pale	Coccidiosis (rare); wheat (fed in place of corn); white corn

COOKED EGGS

Color	Causes
Gray or green surface	Iron in cooking water; overcooked yolk
Green rings	Iron in hen's feed or water
Greenish when scrambled	Egg left too long on a steam table; overcooked egg
Greenish when soft cooked	Fried or served with blueberry pancakes
Yellow rings	Normal layers of yolk

From: *Storey's Guide to Raising Chickens*, 3rd edition

Y

zoning \ Laws regulating or restricting the use of land for a particular purpose, such as raising chickens. Local ordinances may limit or prohibit chicken-keeping activities, including birds bought, sold, traded, shown, shipped, bred, or hatched. Common regulations restrict how many chickens may be kept, how far they must be from the property line, and whether or not roosters are allowed. Information on local zoning ordinances may be obtained from your town or county zoning board and possibly from your county Extension agent.

zoonosis (plural: zoonoses) \

A disease that is transmissible from a chicken (or other animal) to a human. Many of the organisms that cause zoonotic diseases are fairly common in the human environment, whether or not that environment includes chickens. Ordinarily, those pathogens cause no problems. Exceptions are in a person with an impaired immune system or low resistance due to systemic therapy (immunosuppressive or antibacterial

therapy, for example), pregnancy, obesity, diabetes, or a disease unrelated to chickens.

Unfortunately, a chicken carrying zoonotic pathogens may not appear ill but could still cause serious illness in a human. A person can contract a zoonotic disease in many different ways:

- By direct contact with an infected chicken
- By handling contaminated equipment
- Through contact with filth flies or other insects in the environment
- By breathing contaminated dust
- By eating unclean undercooked meat or eggs

Nevertheless, the chance that you will get a disease from your chickens is slim, especially if you routinely follow these commonsense hygienic practices:

- Wash your hands after working with or around your chickens, particularly before eating or drinking anything
- Wear respiratory protection when cleaning your chicken shelter or otherwise stirring up dust
- Wear gloves when handling sick chickens, especially if you have cuts or abrasions on your hands
- Wear coveralls or other protective clothing when working with sick chickens or cleaning their shelter, and launder your clothing immediately afterward
- Most important of all, keep your chickens healthy

zygote \ A fertilized egg before embryonic cell division begins to take place. The word zygote derives from the Greek word *zugōtos*, meaning yoked, in reference to the zygote's receiving genes from both the cock and the hen, which come together to create two sets of chromosomes. The zygote contains all the essential elements for the development of a chick, but they remain encoded as a set of instructions until incubation allows the zygote to develop into an embryo.

Z

QUIZ ANSWERS
BODY TYPE BY BREED

Aseel

Lanshan

Sussex

Sultan

Rosecomb

Spanish

Sumatra

Serama

Brahma

APPENDIX

BREED TRAITS AT A GLANCE

When starting out with chickens or adding new individuals to your flock, learn as much as possible about the differences among breeds. And while you explore chickens' fascinating diversity, think about your own goals, climate, facilities, and personal energy level. Some breeds have innate traits that will help them flourish in your particular setup, others may require extra attention and preparedness, and some simply may not be right for your region, size of operation, or level of experience.

The following chart will help you identify and compare various characteristics of the breeds in this book:

USES. Is the breed esteemed for its eggs, meat, ornamental value, or a combination? Which is most important to you? Is it outstanding in one or more areas, compared to other breeds?

SIZE. Does the breed come in both large and bantam sizes, is it a true bantam, or is it large only?

COMB TYPE. Which of the many types of comb does this breed sport? Some large combs are susceptible to frostbite and those breeds will not be suitable for areas with extreme winters.

CLIMATE TOLERANCE. Is the breed cold-hardy, heat-tolerant, or adaptable to a variety of temperatures? What facilities can you provide during extreme weather?

FORAGING ABILITY. How well will the breed thrive in a pastured situation, compared to others? If you want a low-maintenance breed and have the space, look for an aggressive forager.

EGG SIZE. Where, on the spectrum from Tiny to Extra-Large, do the breed's eggs fit?

SHELL COLOR. Do you want a variety of shell colors in your egg basket or carton?

RATE OF LAY. How productive is this breed, compared to others?

BROODY? Do these hens have a strong innate urge to incubate eggs? Or will they quickly leave the nest after laying an egg, so you can easily collect it?

SKIN COLOR. Most chicken breeds' skin is yellow or white; a few have black skin.

TEMPERAMENT. Will you be able to handle your birds as if they were pets, or is this breed more flighty or aggressive than others?

PLACE OF ORIGIN. Breeds hail from every continent except Antarctica; where did this one originate?

Note: When a box is blank, it means data on that trait is inconclusive.

This chart is just a beginning. As your flock grows you will have a host of opportunities to observe the marvelous variety among chicken breeds and the unique qualities of each individual bird.

BREED TRAITS AT A GLANCE

Breed	Use(s)	Size	Comb Type	Climate Tolerance	Foraging Ability
AMERAUCANA	(eggs, meat)	Large / Bantam	Pea	❄	Good
AMERICAN GAME BANTAM	(ornamental)	Bantam	Single	✳	Aggressive
ANCONA	(eggs) +	Large / Bantam	Single, rose	✳ ❄	Aggressive
ANDALUSIAN	(eggs, ornamental)	Large / Bantam	Single	✳	Aggressive
ARAUCANA	(eggs, ornamental)	Large / Bantam	Pea	❄	Good
ASEEL	(meat, ornamental)	Large / Bantam	Pea	✳ ❄	Good
AUSTRALORP	(eggs, meat)	Large / Bantam	Single	❄	Good
BARNEVELDER	(eggs, meat)	Large / Bantam	Single	❄	Good
BELGIAN BEARDED D'ANVERS	(ornamental)	True Bantam	Rose	✳	
BELGIAN BEARDED D'UCCLE	(ornamental)	True Bantam	Single	✳	
BOOTED BANTAM	(ornamental)	True Bantam	Single	✳	
BRAHMA	(meat)	Large / Bantam	Pea	❄	Good
BUCKEYE	(eggs, meat)	Large / Bantam	Pea	❄	Good
CALIFORNIA GRAY	(eggs, meat)	Large	Single	❄	
CAMPINE	(eggs)	Large / Bantam	Single	❄	Good
CATALANA	(eggs, meat)	Large / Bantam	Single	✳	Good
CHANTECLER	(eggs, meat)	Large / Bantam	Cushion	❄	Good
COCHIN	(meat, ornamental)	Large / Bantam	Single	❄	Good
CORNISH	(meat) + (ornamental)	Large / Bantam	Pea	❄	

● = Eggs 🍗 = Meat 🪶 = Ornamental
+ = Eggs or meat of exceptional value

❄ = Cold-hardy ✳ = Heat-tolerant

Egg Size	Shell Color	Rate of Lay	Broody?	Skin Color	Temperament	Place of Origin
Med–Lg	○	✓	Yes	○	Docile	Chile
Peewee	○		Yes		Varies	USA
Med–Lg	○	✓✓✓	No		Flighty	Italy
Lg	○	✓✓	No		Flighty	Spain
Med–Lg	○	✓	Yes	○	Docile	Chile
Med	○		Yes	○	Docile, some aggressive individuals	India
Lg	○	✓	Yes	○	Docile	Australia
Med–Lg	●	✓	No	○	Docile	Netherlands
Peewee	○		Yes		Docile	Belgium
Peewee	○		Yes		Docile	Belgium
Peewee	○		Yes		Docile	Asia / Netherlands
Med–Lg	○		Yes	○	Docile	India / USA
	○ ○		Yes	○	Docile	Ohio
	○	✓✓✓	Some		Docile	California
Med–Lg	○	✓	No		Docile	Belgium
Med	○	✓	No	○	Flighty	Spain
Lg	○	✓	Yes	○	Flighty	Canada
Med–Lg	○		Yes	○	Docile	Vietnam
Small	○		Yes	○	Docile, some aggressive individuals	England

✓ = Good (150–200 eggs per year)   ✓✓ = Better (200–250 eggs per year)
✓✓✓ = Best (close to 300 eggs per year)

BREED TRAITS AT A GLANCE

Breed	Use(s)	Size	Comb Type	Climate Tolerance	Foraging Ability
CREVECOEUR	Meat, Ornamental	Large / Bantam		Heat-tolerant, Cold-hardy	
CUBALAYA	Ornamental	Large / Bantam	Pea	Heat-tolerant, Cold-hardy	Good
DELAWARE	Eggs, Meat+, Ornamental	Large / Bantam	Single	Cold-hardy	Good
DOMINIQUE	Eggs, Meat, Ornamental	Large / Bantam	Rose	Cold-hardy	Good
DORKING	Eggs, Meat		Single, rose	Cold-hardy	Good
DUTCH BANTAM	Ornamental	True Bantam	Single	Heat-tolerant	
EMPORDANESA	Eggs	Large	Carnation	Heat-tolerant	Aggressive
FAVEROLLE	Eggs, Meat, Ornamental	Large / Bantam	Single	Cold-hardy	
FAYOUMI	Eggs+	Large	Single	Heat-tolerant	Aggressive
HAMBURG	Eggs, Ornamental	Large / Bantam	Rose	Heat-tolerant, Cold-hardy	Good
HEDEMORA	Eggs	Large	Single	Cold-hardy	Aggressive
HOLLAND	Eggs, Meat	Large / Bantam	Single	Cold-hardy	Good
HOUDAN	Eggs, Meat, Ornamental	Large / Bantam	V	Heat-tolerant, Cold-hardy	Aggressive
JAPANESE BANTAM	Ornamental	True Bantam	Single		
JAVA	Eggs, Meat, Ornamental	Large / Bantam	Single		Good
JERSEY GIANT	Meat	Large / Bantam	Single	Cold-hardy	
KRAIENKOPPE	Ornamental	Large / Bantam	Walnut	Cold-hardy	Aggressive
LA FLECHE	Eggs, Meat, Ornamental	Large / Bantam	V	Heat-tolerant	Aggressive
LAKENVELDER	Eggs, Ornamental	Large / Bantam	Single	Heat-tolerant	Good
LANGSHAN	Eggs, Meat, Ornamental	Large / Bantam	Single	Heat-tolerant, Cold-hardy	Good

= Eggs   = Meat   = Ornamental

+ = Eggs or meat of exceptional value

= Cold-hardy   = Heat-tolerant

Egg Size	Shell Color	Rate of Lay	Broody?	Skin Color	Temperament	Place of Origin
Med–Lg	○		No		Flighty	France
Sm–Med	○		Yes		Aggressive	Cuba
Lg–XLg	◔		Yes	○	Docile	Delaware
Med–Lg	◔	✓	Yes	○	Docile	Old World / Haiti
Med–Lg	○		Yes	○	Docile	England
Small	◔	✓	Yes		Flighty	Java
Lg–XLg	●	✓	No		Flighty	Spain
Med–Lg	◔	✓	Yes	○	Docile, some aggressive individuals	France
Small	○	✓✓✓	Later in life		Flighty	Egypt
Sm–Med	○	✓	No		Flighty	Turkey / Netherlands
Medium	●	✓	Some	●	Docile	Sweden
Large	○		No	○	Docile	USA
Sm–Med	○		Some	○	Docile	France
Peewee	◔		Yes		Flighty	Japan
Med–Lg	◔		Yes	○	Docile	Java (unconfirmed)
XLarge	◔		Yes	○	Docile	New Jersey
Med	○		Yes		Flighty	Netherlands
Med–Lg	○		No	○	Flighty	France
Med	○	✓	No		Flighty	Netherlands
Med–Lg	◔	✓	Yes	○	Docile	China

✓ = Good (150–200 eggs per year)   ✓✓ = Better (200–250 eggs per year)
✓✓✓ = Best (close to 300 eggs per year)

BREED TRAITS AT A GLANCE

Breed	Use(s)	Size	Comb Type	Climate Tolerance	Foraging Ability
LEGHORN	Eggs +, Ornamental	Large / Bantam	Single, rose	Heat-tolerant, Cold-hardy	Good
LEGHORN, WHITE	Eggs	Large	Single	Heat-tolerant	
MALAY	Meat	Large / Bantam	Strawberry	Heat-tolerant	Good
MARANS	Eggs, Meat	Large / Bantam	Single	Cold-hardy	Good
MINORCA	Eggs +, Meat	Large / Bantam	Single, rose	Heat-tolerant	Aggressive
MODERN GAME	Ornamental	Large / Bantam	Single	Heat-tolerant	Good
NAKED NECK	Eggs, Meat, Ornamental	Large / Bantam	Single	Heat-tolerant	Good
NANKIN	Ornamental	True Bantam	Single, rose		
NEW HAMPSHIRE	Eggs, Meat	Large / Bantam	Single	Cold-hardy	Good
NORWEGIAN JAERHON	Eggs +	Large / Bantam	Single	Heat-tolerant, Cold-hardy	Good
OLD ENGLISH GAME	Ornamental	Large / Bantam	Single	Heat-tolerant, Cold-hardy	Aggressive
ORLOFF	Eggs, Meat, Ornamental	Large / Bantam	Walnut	Cold-hardy	
ORPINGTON	Eggs, Meat +	Large / Bantam	Single	Cold-hardy	Good
PENEDESENCA	Eggs, Meat	Large	Carnation	Heat-tolerant	Aggressive
PHOENIX	Ornamental	Large / Bantam	Single	Heat-tolerant	
PLYMOUTH ROCK	Eggs, Meat	Large / Bantam	Single	Heat-tolerant, Cold-hardy	Good
PYNCHEON	Ornamental	True Bantam	Single		
POLISH	Eggs, Ornamental	Large / Bantam	V or none	Heat-tolerant	
REDCAP	Eggs, Meat, Ornamental	Large / Bantam	Large rose	Cold-hardy	Aggressive
RHODE ISLAND RED	Eggs, Meat	Large / Bantam	Single, rose	Heat-tolerant, Cold-hardy	Good

● = Eggs   🍗 = Meat   🪶 = Ornamental
+ = Eggs or meat of exceptional value

❄ = Cold-hardy   ☀ = Heat-tolerant

Egg Size	Shell Color	Rate of Lay	Broody?	Skin Color	Temperament	Place of Origin
Med–Lg	○	✓✓	Some		Flighty	Italy
Large	○	✓✓✓	No		Flighty	Italy
Med	○		Yes	○	Flighty	Malay Peninsula
Large	○	✓	Yes	○	Flighty	France
Lg–XLg	○	✓✓✓	No		Flighty	Spain
Small	○		Yes		Aggressive	England
Large	○	✓	Yes	○	Docile	Transylvania
Tiny	○		Yes		Docile	England
Lg–XLg	○		Yes	○	Docile	New Hampshire
Large	○	✓✓✓	No		Flighty	Norway
Small	○		Yes		Aggressive	England
Medium	○	✓	Yes	○	Docile	Persia
Lg–XLg	○	✓	Yes	○	Docile	England
Medium	●	✓	No	○	Flighty	Spain
Sm–Med	○	✓✓	No		Docile	Japan / Germany
Large	○	✓✓	Yes	○	Docile	Massachusetts
Tiny	○		Some	○	Docile	Belgium (unconfirmed)
Med–Lg	○	✓	No		Docile	Netherlands
Medium	○	✓	No	○	Flighty	England
Lg–XLg	○	✓✓	Some	○	Docile, some aggressive individuals	Rhode Island

✓ = Good (150–200 eggs per year)   ✓✓ = Better (200–250 eggs per year)
✓✓✓ = Best (close to 300 eggs per year)

BREED TRAITS AT A GLANCE

Breed	Use(s)	Size	Comb Type	Climate Tolerance	Foraging Ability
RHODE ISLAND WHITE	Eggs, Ornamental	Large / Bantam	Rose	Cold-hardy	Good
ROSECOMB	Ornamental	True Bantam	Large rose	Heat-tolerant, Cold-hardy	
SEBRIGHT	Ornamental	True Bantam	Rose	Heat-tolerant	
SERAMA	Ornamental	True Bantam	Single	Heat-tolerant	
SHAMO	Eggs, Meat + Ornamental	Large / Bantam	Pea, walnut	Heat-tolerant	
SICILIAN BUTTERCUP	Ornamental	Large / Bantam	Buttercup	Heat-tolerant	
SILKIE	Ornamental	Small	Walnut	Heat-tolerant	
SPANISH	Ornamental	Large / Bantam	Large single	Heat-tolerant	Aggressive
SPITZHAUBEN	Ornamental	Large / Bantam	V	Heat-tolerant, Cold-hardy	Aggressive
SULTAN	Ornamental	Large / Bantam	V	Heat-tolerant	
SUMATRA	Ornamental	Large / Bantam	Pea	Heat-tolerant, Cold-hardy	
SUSSEX	Eggs, Meat, Ornamental	Large / Bantam	Single	Cold-hardy	Good
VORWERK	Eggs, Meat, Ornamental	Large / Bantam	Single	Cold-hardy	
WELSUMER	Eggs, Meat, Ornamental	Large / Bantam	Single	Heat-tolerant, Cold-hardy	Aggressive
WYANDOTTE	Eggs, Meat, Ornamental	Large / Bantam	Rose	Cold-hardy	Good
YOKOHAMA	Ornamental	Large / Bantam	Walnut	Heat-tolerant	

= Eggs = Meat = Ornamental
+ = Eggs or meat of exceptional value

= Cold-hardy = Heat-tolerant

Egg Size	Shell Color	Rate of Lay	Broody?	Skin Color	Temperament	Place of Origin
Lg–XLg	○	✓	No	○	Docile	Rhode Island
Peewee	○		No		Docile	England
Tiny	○		No		Flighty	England
Tiny	○		Yes		Docile	Malaysia
Large	○		Yes		Docile, some aggressive individuals	Thailand
Sm–Med	○		No		Flighty	Sicily
Peewee	○		Yes	●	Docile	China
Large	○	✓	No		Flighty	Spain
Medium	○		Yes		Flighty	Switzerland
Small	○		No		Docile	Turkey
Medium	○	✓	Yes	●	Flighty, some aggressive individuals	Sumatra
Large	○	✓	No	○	Docile	England
Med	○		Some			Germany / USA
Med–Lg	●	✓	No	○	Docile	Netherlands
Large	○	✓	Some	○	Docile, some aggressive individuals	Huron lands, USA
Sm–Med	○		Some		Docile	Japan / Germany

✓ = Good (150–200 eggs per year)   ✓✓ = Better (200–250 eggs per year)
✓✓✓ = Best (close to 300 eggs per year)

INDEX

Check first in the alphabetical encyclopedia for the topic you seek; this index provides additional references for topics that appear in more than one listing. Page numbers in **bold font** indicate illustrations and photographs, and *italicized* page numbers indicate charts and tables.

B

back
    butchering and, **44**
    carriage and, 52, **69**
    kinky, 158
    meat breeds and, 175, **175**
    overview of, 24, 232, 282
    roach, 226
bacteria, diseases and, *85*
banding
    of leg, 44, 165
    overview of, 24
    pedigree tracking and, 200
    radio frequency identification and, 220
    spiral, **252**, 253
    of wing, 292, **293**
Bantams, classification of, 59
Bantam Standard, 25, 70, 257
battery brooder, 39
beaks. *See also* Debeaking
    anatomy of, **18**, **19**
    bleaching sequence and, 29–30, *30*
    coloration of, 145
    crossed, 77, **77**
    overview of, 26
    trimming of, 27, **27**
beak trimming, 27
beards, 26–27, **27**, 182, 281
bell waterers, 288, **288**
bird flu, 31, 224, 260
biting flies, 127, 281
blood
    anemia and, 14, 152
    in droppings, 62, 90, 108, *195*
    infections in, 65, 152, 236
    removal of, 32
    on shell, 241
bloom
    anatomy of, 241, **241**
    egg washing and, 104

overview of, 32–33, 98
    shell color and, 242, 243, **243**
blowout, 33, 205, 216, 282
blue chickens, genetics of, *33*
blue eggs. *See under* Eggs
Blue Hen's Chickens, 34
blunting, 257, 258
body shape, age determination and, 8
body type, by breed, 275, **276–277**, 302
Botulism, 16, 36, 296
box brooder, 39
breast
    anatomy of, **18**, **19**, 37
    butchering and, **44**
    dark meat, light meat and, 81
    defeathered patch on, 42
    meat breeds and, 175, **175**
    molting sequence and, 181, **181**
breeds
    by body type, 275, **276–277**, **302**
    summary of traits of, *303–311*
bristles, 115, **115**
broilers
    controlled lighting for, 71
    Cornish Rock, 74–75
    droppings output per chicken, *91*
    feather sexing and, 117, **117**
    feed guidelines for, 118, 120–121, 216, *217*
    feeding of, 118, 120–121
    grain-fed, 135
    overview of, 38–39
    overview of meat class, 81, 176
    pasture raised, 197, 237
    temperature and, 267
bronchitis, shells and, 242

butchering, **44**, **90**, 223
buttercup comb, 43, **67**

C

cackle, 9–10, 45
calcareous layer, 241, **241**
calcium
    dehydration and, 105
    overview of, 46
    phosphorus and, 204
    shell strength and, 102, 242, 244, 284
    supplementing, 17, 24, 152, 244, 284
calcium carbonate, 17, 46–47, 136
calcium grit, 46–47, **46**, 136
Caldwell, Jonathan, 34
campylobacteriosis, 48, 108
candling
    air cell and, 9, 150
    blood spot and, 32
    egg freshness and, 99, **99**
    egg quality and, 100
    overview of, 49, **49**
cannibalism
    eggs and, 98–99
    feathers and, 116
    forms of, *50*
    pecking order and, 200
    salt and, 51, 232
    signs of, *211*
    toe picking as, 269
capillary worms, *153*
capon, 51, **51**, 176
carnation comb, 51, **51**, 59, **67**
carotenoids, 297
carriers, diseases and, 86
carrying chickens, 52, **52**
castration, 51, 176
caution call, 9
cecal worms, 54, *153*, 231

diapers, 56, **56**, 146

diatomaceous earth, 84, 93

digestive anatomy, **60**

diseases. *See also Specific diseases*
  biosecurity and, 29
  of blood, 14, 236, 241
  droppings color and, 89–90
  eggs and, 242
  enteric, 108, 215
  enzootic, epizootic, 108
  fertility and, 124, 125
  flies and, 127
  housing and, 10, 45, 126
  immunity and, 15, 41, 148,
    194, 236, 295
  laxative flush and, 162
  molting and, 181
  opportunistic, 190
  overview of, 84–86
  reportable, 224
  respiratory, 13, 58, 220, 222–
    223, 224–225
  summary of causes of, *85*
  toxins and, 7, 16, 36, 270,
    272–273
  transmission of, 52, 178, 187,
    197, 227, 281, 288
  vaccination and, 280
  zoonoses, 31, 301

disinfectant, 86, 207–208, 233

dolomitic limestone, ground,
  14, 46

dominant genes, 10, 34, 87, 131,
  133, 218, 239

drafts, 41, 88–89, 261

droppings. *See also* Manure
  blood in, 62, 90, 108, *195*
  cecal, 54, 91
  disease and, 31, 48, 65,
    85–86, 108, 117, *195*, 232,
    251
  fecal testing and, 117

fertilizer value of, *172*
  overview of, 89–91, *91*

drowning, 42, 288–289

dyeing, of embryos, **107**

E

ear, anatomy of, **95**

ear lobe, *30*, 92, 95, **95**, 243

earthworms, *153*, 167, 217, 266,
  295

eggs. *See also* Yolk
  air cell depth and, 9, *9*
  anatomy of, **54**, **104**
  antibodies in, 15
  avidin and, 23
  calcium and, 46
  candling of, 49, **49**
  cannibalism and, *50*, 98–99
  cholesterol and, 58
  colored, 11, 34, 58, 66, 95,
    173, 243, 245, *245*
  development of, **98**
  draggy hatching and, 89
  ear lobe color and, 95
  fake, 185
  floating of, **99**
  genetic determination of
    color of, 34
  heritability of traits in, *143*
  identifying predators of, *213*
  laying rates of, *165*, 222
  shape of, 100, **100**
  shell and, 241–244, **241**, 245
  size of, *101*
  wind, **292**
  yolkless, 63, 94, **292**, 299

Eimeria spp., 61, 62

electrolytes, 105, *105*, 141

embryos, 49, **106**, **107**

enteritis, 48, 108, 185

epsom salts, laxative flush and,
  162, 163

Escherichia coli, 65, 108

estrogen, 100, 145

extension offices, 73, 153, 224,
  280, 301

eye ring, bleaching sequence
  and, *30*

F

face, coloration of, **111**

FARAD. *See* Food Animal
  Residue Avoidance and
  Depletion Program
  (FARAD)

fear, of chickens and parts, 10

feathers
  anatomy of, **19**, **114**, **115**
  axial, **23**, **115**
  brood patches and, 42
  cannibalism and, *50*, 116
  clubbing down and, 61
  coloration of, 136, 206
  crests and, 76
  debraining and, 82
  fertility and, **115**, 125
  flight, **127**
  frizzled, 130, **130**
  hackle, 137, **137**
  lack of, 116, **116**, 180, 182, 183
  molting and, 180–181
  patterns of, 25, **25**, **88**, **96**,
    112, 159, 178, 202, 251,
    260, **261**
  pinfeathers, 205
  positions of, **19**, **115**
  primary, 214, **215**
  secondary, 236
  sexing and, **117**
  tracts and, **117**
  tufts and, **95**
  wing, 293–294, **294**

feeding/feeders, 118–123, **118**,
  119, **120**, **123**, **274**

lime sulfur and, 167

northern fowl, 187, *195*

overview of, 7, 180, 194

red, *195*, 223

scaly leg, *195*, 234

signs of common, *195*

moisture, 41. *See also* Humidity

molasses, 162, 163

mold, 7, *85*, 182

molting, 130, 180–181, **181**, 261, 294, **294**

mortality rate, 81, 181

muscles, 8, 81, 125

N

nakedness, 116, **116**, 180, 182, 183

National Institute of Food and Agriculture, 72

nesting boxes, guidelines for, *184*

neuraminidase proteins, bird flu and, 31

nitrogen, chicken manure and, 68–69, *172*

nutrition

diseases and, *85*

feeds and, 24, 38, 56, 66–67, 118, 120, 123, 202

luster and, 70, 169

molting and, 130

pasture and, 196–197

problems of, 50, 85, 98, 152, 189, 208, 243, 261, 267, 268, 296

shell color and, 243

supplements and, 263, 270

O

omega-3 fatty acids, 126, 189–190, 197

operant conditioning, 28, 190

organic, 183, 184, 190–191

ornamental breeds, 13, 167, 191, 192

ova, **98**

ovaries, 79, 193, **193**

P

parasites. *See also Specific parasites and diseases*

diseases and, *85*, 108

fertility and, 125

hosts of, 127, 128–129, 145, 153, *153*, 166–167, 295

overview of, 194, 265, 295

signs of common, *195*

treating/preventing, 83–84, 93, 153, 197, 229–230, 280

pasting, pasty butt, 196

pea comb, *67*, 198, **198**

Pearson's square, 121, 122, 199

pecking order, 64, 198, 200

phosphate, 14, 68, 82, *172*, 204

pigment

bleaching and, 29–30

blues and, 34

shell color and, 33, 242, 243, **243**

skin color and, 29–30

xanthophyll, 297, 299

plants, toxic, 208, 270, **272–273**

plumules, **114**, 115

pneumatic bones, air sacs and, 9

poisoning

aflatoxins and, 7

botulism and, 36

chemical, *85*, 187

enterotoxins and, 108

laxative flush and, 162

mycotoxins and, 182

overview of, 207–208

rodenticides and, 227

salt and, 233

septicemia, 236

Staphylococcus aureus and, 20

from toxic plants, 208, 270, **272–273**

potash, 68, 69, *172*

predators, 41, *211–213*, 226–227

probiotics, 62, 68, 196, 214–215

prolactin, 42

protein

amino acids and, 13, 108

overview of, 15, 216–217

Pearson's square and, 199

requirements for, 120–123, *217*, 225

sources of, 217

pterylae, 117, **117**

pullets, 59, 72, 118, 122, 218, 259

Punnett Square, 218, **218**

R

Rare Breeds Canada, 107, 221

rate of lay, *164*, 222, 224, 287, *304–311*

recessive genes, 10, 133, 166, 218, 223, 239, 267

recycling, of egg shells, 46

respiratory disease, 13–14, 58, 220, 222–225. *See also Specific diseases*

Revolutionary War, 34

riboflavin (vitamin B_2), 61, 226, 284

roaster, overview of meat class, 176

rose comb, *67*, 153, **228**, 229, 252

rotation, **230**

rump, 125, 133, 231, 278. *See also* Araucana

PHOTO AND ILLUSTRATION CREDITS

Interior photography by © Adam Mastoon: 61, 111, 312, and 206 right (all); Alex Ford/Wikimedia Commons: 151; © Andreas Pistolesi/Getty Images: 64 bottom; © Cackle Hatchery: 223; © catnap72/iStockphoto.com: 209; © Christopher Taylor/Alamy: 89; © David Chapman/Alamy: 58; © Eric Futran - Chefshots/Foodpix/Getty Images: 126; Evelyn Simale/Wikimedia Commons: 197; FOODCOLLECTION/ageFotostock: 123; © Gail Damerow: 29 middle, 46, 103, 136, 140, 189, 201, 235, 242, 288, and 292; © Gail Thuner: 109; © Graphic Science/Alamy: 144; © Isselee/Dreamstime.com: 231; © John Valls: 57; © Krys Bailey/Alamy: 93; © Marcelino Vilaubi: 34, 59, 95, 101, 110, 118–120, 139, 196, 289, and 293; Marcelino Vilaubi: 29 bottom and 243; © Michael Clinton and The Roslin Institute: 29 top; © Mommamoon/Dreamstime: 88; © Moshe Miner/GPO/Getty Images: 116; © Nathan Luke/Alamy: 184; © Nature PL/SuperStock: 206 left; © Petko Danov/iStockphoto.com: 64 top; © Pixitive/iStockphoto.com: 97; © poco-bw/iStockphoto.com: 170; © Ryan Rodrick Beller/Alamy: 45; © Dr. Jacqueline Jacob, University of Kentucky: 117; Wanda Zwart/Wikimedia Commons: 182; Wikimedia Commons: 79 and 185
Cover and interior illustrations by © Bethany Caskey, except for Beverly Duncan: 272 (row 2, left and center and row 3, left), 273 (row 1, center); © Elayne Sears: 60, 137, 175, 221, 272 (row 3, center), 273 (row 2, center and row 3, right); © Fotosearch/ageFotostock: 169; © James Robins: 114, 115 top, 127, and 240; Jeffrey C. Domm: 2 (egg, bottom right), 54, and 104; Louise Riotte: 272 (row 1, left and row 3, right), 273 (row 1, right and row 2, left); Regina Hughes: 272 (row 1, right and row 2, right), 273 (row 1, left; row 2, right; row 3, left and center)

COMPLETE YOUR CHICKEN EDUCATION WITH
MORE STOREY BOOKS

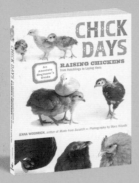

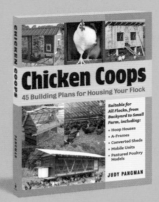

**Jenna Woginrich,
photography by Mars Vilaubi**
Absolute beginners will delight in this photographic guide chronicling the journey of three chickens from newly hatched to fully grown, highlighting all the must-know details about chicken behavior, feeding, housing, hygiene, and health care.

Judy Pangman
This collection of 45 hen hideaways promises to spark your imagination and inspire you to begin building. Everything you need is detailed in the basic plans, elevation drawings, and building ideas for these original coops.

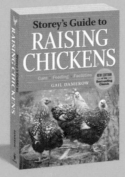

Gail Damerow
This comprehensive volume addresses every aspect of chicken health, including nutrition, diseases, parasites, reproductive issues, immune health and much more, with detailed solutions for problems you may encounter, including those specific to raising birds in the city.

Gail Damerow
Here's all the information you need to raise happy, healthy chickens. Learn about everything from selecting breeds and building coops to incubating eggs, hatching chicks, keeping your birds healthy and safe from predators, and much more.